EXPERIMENTAL AND MODELING STUDIES OF HORIZONTAL SUBSURFACE FLOW CONSTRUCTED WETLANDS TREATING DOMESTIC WASTEWATER

NJENGA MBURU

Thesis committee

Promotor
Prof. Dr P.N.L. Lens
Professor of Environmental Biotechnology
UNESCO-IHE, Delft

Co-promotors
Prof. Dr D.P.L. Rousseau
Professor of Environmental Sciences
Ghent University, Kortrijk, Belgium

Dr J. J.A. van Bruggen
Senior lecturer in Microbiology
UNESCO-IHE, Delft

Other members
Prof. Dr G. Zeeman, Wageningen University
Dr T. Okia Okurut, National Environment Management Authority, Kampala, Uganda
Dipl.-ing. Dr G. Langergraber, University of Natural Resources and Life Sciences, Vienna, Austria
Dr N. Fonder, Namur Provincial Technique Service, Belgium

This research was conducted under the auspices of the SENSE Research School for Socio-Economic and Natural Sciences of the Environment

Experimental and Modeling Studies of Horizontal Subsurface Flow Constructed Wetlands Treating Domestic Wastewater

Thesis
submitted in fulfilment of the requirements of
the Academic Board of Wageningen University and
the Academic Board of the UNESCO-IHE Institute for Water Education
for the degree of doctor
to be defended in public
on Friday, 29 November 2013 at 1.30 p.m.
in Delft, The Netherlands

by

NJENGA MBURU
Born in Mombasa, Kenya

CRC Press/Balkema is an imprint of the Taylor & Francis Group, an informa business

© 2013, Njenga Mburu

Published by:
CRC Press/Balkema
PO Box 11320, 2301 EH Leiden, The Netherlands
e-mail: Pub.NL@taylorandfrancis.com
www.crcpress.com – www.taylorandfrancis.com

ISBN 978-1-138-01552-4 (Taylor & Francis Group)
ISBN 978-94-6173-768-7 (Wageningen University)

Dedication

To David, Mercy and Beatrice

Acknowledgements

The writing of this dissertation has been made possible by the guidance of my promoter and mentors, help from local supervisor, financial assistance from the Dutch Government and support from my wife.

I firstly would like to acknowledge my promoter, Professor Dr. ir. P.N.L. Lens of UNESCO-IHE Institute for Water Education. It was a pleasure to learn and work under his supervision. He significantly helped me to better structure and sharpen my writing of the manuscripts for the peer review process, besides bringing the much required thrust towards the conclusion of the study.

This work was largely shaped through the guidance of my mentor Professor Dr. ir. D.P.L. Rousseau of Ghent University, department of Industrial Biological Sciences. He is the mind behind the modeling study within the framework of Constructed Wetland Model No. 1. The multiple discussions and E-mail exchanges we had in the course of this study undeniably contributed to bringing this work this far. I would also like to thank my mentor, Senior lecturer, Dr. J.J. van Bruggen of UNESCO-IHE Institute for Water Education, through whom I made the first contact with UNESCO-IHE institute and has helped in sustaining this study. I'm grateful for the trips my mentors made to Kenya.

Special thank to Dr. Esther Llorens of University of Girona, Spain for the training she provided on the modeling software. Indeed, thanks too to Professor Dr. J. García of Universitat Politècnica de Catalunya-BarcelonaTech for hosting me at his department and organizing the memorable tours of Constructed Wetland facilities in Barcelona. Many thanks to my local supervisor Professor Dr. G.M. Thumbi of the Technical University of Kenya. He was helpful with the local administrative issues and logistics during the study.

I'm extremely grateful to the Netherlands Government for providing financial assistance through the Netherlands Fellowship Program. I would also wish to extend my sincere thanks to Masinde Muliro University of Science and Technology, Kakamega, Kenya for granting me a study leave to pursue the study.

Great and profound appreciation to my lovely wife Catherine, my son David and daughters Mercy and Beatrice; I deeply appreciate your patience. Finally, I acknowledge all those who supported me but whose names have not featured on this page.

Contents

Chapter 1: General introduction

There are two fundamental reasons for treating wastewater: to prevent pollution, thereby protecting the environment and protecting public health by safeguarding water supplies and preventing the spread of water-borne diseases (Gray, 2004). Throughout history wastewater management has presented people and governments with far reaching technical, economical and political challenges that has necessitated the development of wastewater management strategies in different periods and cultures (Lofrano and Brown, 2010). In the 21st century it can be said that technologies to treat wastewater are now well established and are capable of producing almost any degree of purification.

Joining the "suite" of wastewater treatment technologies in the fifties of the past century were the Constructed Wetlands (CWs). Research in the 1950's by Dr. Kathy Seidel of the Max Planck Institute in Germany, brought recognition to constructed wetlands as a viable wastewater treatment technology (Vymazal, 2005). Exponential growth in the application of CWs has been experienced in the past three decades. This wastewater treatment technology has emerged as a sustainable, environmentally friendly solution in many countries across the world (Kadlec and Wallace, 2009). Because of the ability of CWs to recycle, transfer and/or immobilize a wide range of potential contaminants, there are an ever expanding number of applications to treat different types of polluted water (Greenway, 2007). Indeed besides domestic and municipal application, the use of constructed wetlands has spread to many other fields including treatment of industrial effluent, special wastewater (e.g. from hospitals, acid mine drainage), agricultural effluent, landfill leachate, road runoff and sludge consolidation (Vymazal, 2010b). In the developing countries, the use of CWs for water purification is particularly valuable and exploitable for the protection of water quality in catchments, rivers and lakes (Denny, 1997).

1.1 Inadequate sanitation in developing countries

It is still a concern that safe management of wastewater is not universal. Sanitation targets have joined the millennium developments goals (Mara et al., 2007; WSSCC, 2004), articulating an international commitment of halving the proportion of people without access to safe water and sanitation by 2015 through an integrated approach to sanitation, water supply, and hygiene promotion. The bulk of the world population without adequate sanitation

2

lives in the developing countries (WHO and UNICEF, 2000) where the evacuation and treatment of wastewater from human settlements is not satisfactory. The country of Kenya is no exception in this regard (GOK, 2005). The combination of rapid population growth, industrialization, expansion of agriculture and associated urbanization has increased wastewater volumes. The main wastewater treatment and sanitation facilities remain concentrated in the formally planned urban areas, though most people live in rural areas and the urban informal settlements (GOK, 2005). The differences in access to adequate sanitation between urban and rural environments still persist with urban areas served better than the rural areas. Outbreaks of waterborne diseases in addition to the eutrophication of surface water resources is common place (Odada et al., 2004). Approximately 80% of the outpatient hospital attendance in Kenya is due to preventable diseases while 50% of these are water, sanitation and hygiene related (Njuguna and Muruka, 2011).

Generally the reason for the lack of ample wastewater treatment in developing countries is financial (Muga and Mihelcic, 2008) - the large investments required to construct, maintain or upgrade wastewater treatment facilities - due to a focus on high cost technology; but it is also due to a lack of keeping abreast with the advances in wastewater technologies and the application of low-cost wastewater treatment technologies (Tsagarakis et al., 2001). The cost of wastewater treatment and pollution control is high, and rising annually, not only due to inflation but to the continuous increase in environmental quality that is expected (Gray, 2004). Thus, the main issue surrounding the selection of a given wastewater treatment process lies in deciding which is the most appropriate and applicable technology for a particular social, political and economic environment (Tsagarakis et al., 2001). For developing countries, short-term and medium-term solutions for wastewater management lie in research focused on the best technologies suited to the local conditions and the use of cheap and robust wastewater treatment technologies in tandem with the economic realities in these low to middle income countries (Okurut, 2000; Tsagarakis et al., 2001). A precedent where a number of cheap wastewater treatment technologies such as constructed wetlands have been researched on and applied is in developed countries including the USA and EU (Kadlec and Wallace, 2009). Indeed they have been found useful in minimizing environmental pollution by removing oxygen depleting substances, the major nutrients (nitrogen and phosphorus), as well as pathogenic microorganisms from wastewater (Vymazal, 2010a).

In Kenya, the long-term national planning strategy, officially known as *Kenya Vision 2030* aspires for a country firmly interconnected through a network of roads, railways, ports, airports, water and sanitation facilities, and telecommunications (GOK, 2007). The 2030 vision for water and sanitation is to ensure that improved water and sanitation are available and accessible to all. In line with the millennium development goals, this is to be achieved by (a) the rehabilitation and expansion of urban and rural water supply and sanitation in the key satellite towns, and (b) special attention to support projects that will contribute to sustainable water management and good sanitation through increased investments in research and experimental development to generate innovative scientific solutions and new technologies.

In this work, the focus is on the application of constructed wetland wastewater treatment technologies for the treatment of domestic wastewater.

1.2   Treatment wetlands: option for wastewater management

Constructed wetlands are generally designed or engineered to mimic and optimize many of the wastewater treatment conditions and/or processes that occur in natural wetlands (Langergraber et al., 2009a; Vymazal and Kröpfelová, 2009). It presents an important alternative to conventional wastewater management technologies in developing countries (Diemont, 2006; Kivaisi, 2001) for the following reasons:

(a)   In developing countries the emphasis now is on low-cost, high-performance, sustainable wastewater treatment systems (Muga and Mihelcic, 2008). Most of these systems would be 'natural' systems - so called because they do not require any electromechanical power input - with simple configuration together with low energy requirements and operating cost. The replication of centralized, highly engineered wastewater management systems has not been successful or possible everywhere. Often they are installed only in the urban and semi urban areas in developing countries. In the event the municipal authorities are not able to meet the costs of energy, chemicals and skilled labor to run the facilities, the treatment plants deteriorate and eventually break down some years after commissioning.

(b)   The configuration can be adapted to a form that is suitable for small communities such as individual households and farms. This directly satisfies the smaller rural units and people without economic resources to offset the high per capita costs of sewage collection and treatment inherent in conventional wastewater treatment technologies. Eventually such technologies benefit as many people as possible while being adaptable to the changing needs of the community which are important criteria for a technology to be successful in developing countries (Haberl, 1999).

(c)   In an era where there is growing concern of the local and global impact of our current environmental management strategies, and the need to reduce sanitation problems, disease, and poverty, there is a greater need to develop more environmentally responsible, appropriate wastewater treatment technologies whose performance is balanced by environmental, economic, and societal sustainability (Muga and Mihelcic, 2008).

1.3   The horizontal subsurface flow constructed wetland (HSSF-CW)

Among the treatment wetlands, the HSSF-CWs are a widely applied concept (Vymazal, 2005). Pre-treated (to prevent clogging) wastewater flows horizontally (i.e. the inlet and outlet horizontally opposed) through the artificial filter bed of porous media, usually consisting of a matrix of sand or gravel and the macrophyte roots and rhizomes. This matrix is colonised by a layer of attached microorganisms that forms a so-called biofilm. Purification is achieved by a variety of physical, chemical and (micro)biological processes, such as sedimentation, filtration, precipitation, sorption, plant uptake, and microbial decomposition (García et al., 2010; Kadlec and Wallace, 2009).

The subsurface flow regime has a major impact on the physicochemical conditions that develop within the HSSF-CW treatment reactor and the types of vegetation that can be grown in this type of wetland (Fonder, 2010). The HSSF-CW is typically planted with sessile emergent vegetation (Fig.1.1). The HSSF-CW have the advantage of a lower footprint or area requirement (because the contact area of water, microorganisms and substrate is large), lower risk of spread of vector disease and odour, compared to the surface flow type wetland in which the majority of wastewater flows exposed to the atmosphere.

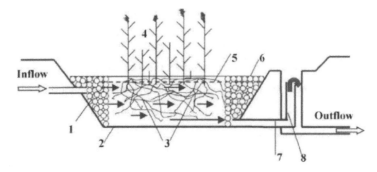

Fig. 1.1. Schematic representation of a constructed wetland with horizontal sub-surface flow. 1, distribution zone filled with large stones; 2, impermeable liner; 3, filtration medium (gravel, crushed rock); 4, vegetation; 5, water level in the bed; 6, collection zone filled with large stones; 7, collection drainage pipe; 8, outlet structure for maintaining of water level in the bed. The arrows indicate only a general flow pattern (After Vymazal, 2005).

HSSF-CW are typically used to treat primary or secondary treated sewage prior to discharge. In general, HSSF-CW have been utilized for relatively smaller flow rates than the surface flow constructed wetlands, mainly because of cost and hydraulic limitations associated with flow through the porous media. These systems are capable of operation under colder conditions than the surface flow systems because of the ability to insulate the top surface and the thermal buffering provided by the substrate.

1.4   Mechanistic modeling of horizontal subsurface flow constructed wetlands

Despite there being much experience available in constructing and operating CWs, it is still difficult to evaluate and improve their existing design criteria. Some aspects of their performance are still unknown because the degradation of wastewater contaminants within these systems takes place through a number of physical, chemical and biological processes that occur simultaneously. They are still often looked at as "black box" technology in which the wastewater is purified and only simulated using various degradation models, mostly based on input/output data (Rousseau et al., 2004; Wynn and Liehr, 2001). Indeed, CWs are known to be complex systems, the behaviour of which depends on both external (e.g. flow-rate, wastewater composition and temperature) and internal (e.g. bacteria growth and development) factors (Samsó and Garcia, 2013). Further, there is a lack of direct observational methodologies or the required equipment may be sophisticated and relatively expensive. Nevertheless, for the sound design and optimization of CWs, the following questions need resolution:

1. How do the microbial communities in the constructed wetland interact as far as the cycling of carbon, nitrogen and sulphur is concerned? What is the functional composition of the different bacterial groups?

2. What is the contribution of the aerobic, anoxic and anaerobic microbial pathways in horizontal subsurface flow constructed wetlands?

3. Are sorption processes significant for delaying COD and ammonium release in subsurface flow wetlands?

4. What is the magnitude of root zone aeration?

Mechanistic numerical modeling and simulation has become an efficient and elegant way to obtain a better understanding of the performance of wastewater treatment systems (Llorens et al., 2011). During the last decade, there has been a wide interest in the understanding of complex "constructed wetland" systems, including the development of numerical process-based models describing these systems (Langergraber and Šimůnek, 2012). Process-based or mechanistic models that describe both water flow and reactions in CWs in detail are potentially faster and more economical tools for the qualitative and quantitative interpretation of the complex CWs processes and performance.

Different formulations for the reaction/ biokinetic model have been developed by different authors, mainly because the models have been developed for different applications (Langergraber et al., 2009a). In 2009, the Constructed Wetland Model N°1 (CWM1) was introduced. CWM1 describes all relevant aerobic, anoxic and anaerobic biokinetic processes occurring in subsurface flow CWs required to predict effluent concentrations of organic matter, nitrogen and sulphur (Langergraber et al., 2009b). With the publication of CWM1, it is supposed that such as for the International Water Association (IWA) activated sludge models, mechanistic modeling will become unified and popular as a supporting tool for design and operation of subsurface flow CWs. However, as the process-based modeling of constructed wetlands is still in the early stages of development (Langergraber and Šimůnek, 2012), several requirements still remain before the good simulation of CWs is achieved and the process-based models can be used for evaluating design criteria, for example:

- Other processes besides microbial transformation and degradation have to be considered for the formulation of a full model for CWs, i.e. the influence of plants (growth, decay, decomposition, nutrient uptake, root oxygen release, etc.), transport of particles/suspended matter and the description of clogging processes, adsorption and desorption processes, as well as physical re-aeration on the treatment performance of CWs. The presence and species of wetland plants have been shown to influence disposal of pollutants in CWs (Dhote and Dixit, 2009; Gagnon et al., 2007), while clogging of the subsurface flow in CWs is still one of the main, often-occurring, operational problems (Knowles et al., 2011). The inclusion of such a process in CW models would allow prediction of the failure of subsurface flow CWs due to clogging.

- The implementation of biokinetic reactions and other CW processes in suitable software to solve the differential equations for dynamic simulations of reactive transport in subsurface flow constructed wetlands. Simulation software that allows for spatially and temporally resolved process specific mass balances for reactive species and possible simulation of fixed biomass would support the understanding and prediction of biochemical transformations and degradation processes in constructed wetlands. Further, this would facilitate the validation of CW process-based models in a wide range of conditions encountered in the design and evaluation of constructed wetland systems.

Some of these issues were tackled in this thesis.

1.4   Problem statements and thesis outline

For a sound conceptualization and design of efficient and economic constructed wetland systems, there is still a need for an improved understanding of the internal processes involved in the transformation and degradation of carbon, nitrogen and sulphur within the horizontal subsurface constructed wetland for the treatment of domestic wastewater. Further, performance data and information that can guide design and operation of CWs under tropical conditions is scarce. These issues are addressed in the thesis in two parts: (a) through information and data collection and analysis from the literature and experimental set-up (Chapters 2-4), in combination with (b) mechanistic numerical modeling (Chapters 5-7).

In the first part, Chapter 2 presents a literature survey describing the application and performance of the tropical macrophyte *Cyperus papyrus* in wetlands for wastewater treatment. Performance data from the literature and information on growth, productivity and harvesting of *Cyperus papyrus* is collated. The *Cyperus papyrus* macrophyte was chosen as it is among the most productive plants in wetlands besides being the dominant species of many swamps in East and Central Africa. The performance of a HSSF-CW treating primary effluent of domestic wastewater under tropical conditions is evaluated in Chapter 3. The removal of organic matter (COD and BOD_5), suspended solids and nutrients is assessed, together with the influence of the *Cyperus papyrus* macrophyte. Chapter 4 presents a comparative study of the performance and costs based on a full scale installation of a secondary facultative pond and a pilot scale horizontal subsurface flow constructed wetland. The constructed wetland systems have not yet found widespread use in developing countries due to a lack of or little available direct comparative information between constructed wetlands and other locally applied low cost technologies. Such a comparison is intended to offer technical and economic insights that would simplify the selection processes.

In the second part of this thesis, process-based numerical modeling is employed to support the understanding and prediction of the complex biochemical transformation and degradation processes in constructed wetlands. Chapter 5 describes the simulation of reactive transport based on data from the pilot scale constructed wetland described in chapter 3, using the 2D-CWM1-RETRASO model. The 2D simulation model allows the prediction of the effluent concentrations and comparison of those reaction rates involved in the organic matter degradation along the length and depth. In Chapter 6 the results of implementation of the biokinetic model CWM1 in the AQUASIM software for identification and simulation for aquatic systems is discussed. Data from a set of batch-operated constructed wetland mesocosms operated under a range of temperatures is used for calibration and validation. The conversion of COD, NH_4^+-N and SO_4^{2-}-S in batch-operated constructed wetland mesocosms is simulated. The effect of adsorption on COD and NH_4^+-N is evaluated. The impact of temperature and presence of plants on the bacterial concentration is elucidated. Chapter 7 presents an extension of the work in Chapter 6, which involved incorporating a biofilm modeling approach based on CWM1-AQUASIM. This was to elucidate the biofilm growth dynamics in a multispecies bacterial-biofilm which is essential to the design conceptualization of treatment processes for constructed wetlands. Chapter 8 presents a general discussion based on the results from the different chapters. The future outlook of the constructed wetland wastewater treatment technology and its merits as a viable alternative to conventional treatment of domestic wastewater in developing countries is discussed.

1.5   References

Denny, P., 1997. Implementation of constructed wetlands in developing countries. Water Science and Technology 35(5) 27-34.

Dhote, S., Dixit, S., 2009. Water quality improvement through macrophytes—a review. Environmental Monitoring and Assessment 152(1) 149-153.

Diemont, S.A.W., 2006. Mosquito larvae density and pollutant removal in tropical wetland treatment systems in Honduras. Environment International 32(3) 332-341.

Fonder, N., 2010. Hydraulic and removal efficiencies of horizontal flow treatment wetlands (PhD thesis in English). Gembloux, Belgium ULG, Gembloux Agro-Bio Tech 191p.

Gagnon, V., Chazarenc, F., Comeau, Y., Brisson, J., 2007. Influence of macrophyte species on microbial density and activity in constructed wetlands. Water Science and Technology 56(3) 249-254.

García, J., Rousseau, D.P.L., MoratÓ, J., Lesage, E.L.S., Matamoros, V., Bayona, J.M., 2010. Contaminant Removal Processes in Subsurface-Flow Constructed Wetlands: A Review. Critical Reviews in Environmental Science and Technology 40(7) 561-661.

GOK, 2005. MDGs Status report for Kenya 2005. Goverment of Kenya, Ministry of Planning and National Development in partnership with UNDP,Kenya ,and the Government of Finland.

GOK, 2007. Kenya vision 2030. Government of Kenya.

Gray, N.F., 2004. Biology of wastewater treatment. Imperial College Press 57 Shelton Street Covent Garden London WC2H 9HE 2nd Edition.

Greenway, M., 2007. The Role of Macrophytes in Nutrient Removal using Constructed Wetlands Environmental Bioremediation Technologies, In: Singh, S., Tripathi, R. (Eds.). Springer Berlin Heidelberg, pp. 331-351.

Haberl, R., 1999. Constructed wetlands: A chance to solve wastewater problems in developing countries. Water Science and Technology 40(3) 11-17.

Kadlec, Wallace, S., 2009. Treatment wetlands. 2nd ed. Boca Raton, Fla: CRC Press, 1048 pp.

Kivaisi, A.K., 2001. The potential for constructed wetlands for wastewater treatment and reuse in developing countries: a review. Ecological Engineering 16(4) 545-560.

Knowles, P., Dotro, G., Nivala, J., García, J., 2011. Clogging in subsurface-flow treatment wetlands: Occurrence and contributing factors. Ecological Engineering 37(2) 99-112.

Langergraber, G., Giraldi, D., Mena, J., Meyer, D., Pena, M., Toscano, A., Brovelli, A., Korkusuz, E.A., 2009a. Recent developments in numerical modelling of subsurface flow constructed wetlands. Science of the Total Environment 407(13) 3931-3943.

Langergraber, G., Rousseau, D.P.L., Garcia, J., Mena, J., 2009b. CWM1: a general model to describe biokinetic processes in subsurface flow constructed wetlands. Water Science and Technology 59(9) 1687-1697.

Langergraber, G., Šimůnek, J., 2012. Reactive Transport Modeling of Subsurface Flow Constructed Wetlands Using the HYDRUS Wetland Module. Vadose Zone Journal 11(2).

Llorens, Saaltink, M.W., Poch, M., García, J., 2011. Bacterial transformation and biodegradation processes simulation in horizontal subsurface flow constructed wetlands using CWM1-RETRASO. Bioresource Technology 102(2) 928-936.

Lofrano, G., Brown, J., 2010. Wastewater management through the ages: A history of mankind. Science of the Total Environment 408(22) 5254-5264.

Mara, D., Drangert, J.-O., Anh, N.V., Tonderski, A., Gulyas, H., Tonderski, K., 2007. Selection of sustainable sanitation arrangements. Water Policy 9(3) 305-318.

Muga, H.E., Mihelcic, J.R., 2008. Sustainability of wastewater treatment technologies. Journal of Environmental Management 88(3) 437-447.

Njuguna, J., Muruka, C., 2011. Diarrhoea and malnutrition among children in a Kenyan district: a Correlational study. JRuralTropPublicHealth 10 35-38.

Odada, E.O., D. O. Olago, K. Kulindwa, Ntiba, M., Wandiga, S., 2004. Mitigation of Environmental Problems in Lake Victoria, East Africa: Causal Chain and Policy Options Analyses. Ambio 33(1-2).

Okurut, T.O., 2000. A Pilot Study on Municipal Wastewater Treatment Using a Constructed Wetland in Uganda. PhD dissertation, UNESCO-IHE, Institute for Water Education, Delft, The Netherlands.

Rousseau, D.P.L., Vanrolleghem, P.A., De Pauw, N., 2004. Model-based design of horizontal subsurface flow constructed treatment wetlands: a review. Water Research 38(6) 1484-1493.

Samsó, R., Garcia, J., 2013. BIO_PORE, a mathematical model to simulate biofilm growth and water quality improvement in porous media: Application and calibration for constructed wetlands. Ecological Engineering 54(0) 116-127.

Tsagarakis, K.P., Mara, D.D., Angelakis, A.N., 2001. Wastewater management in Greece: experience and lessons for developing countries. Water Science and Technology 44(6) 163-172.

Vymazal, 2005. Horizontal sub-surface flow and hybrid constructed wetlands systems for wastewater treatment. Ecological Engineering 25 478–490.

Vymazal, 2010a. Constructed Wetlands for Wastewater Treatment. Water Air and Soil Pollution 2 530-549.

Vymazal, Kröpfelová, L., 2009. Removal of organics in constructed wetlands with horizontal sub-surface flow: A review of the field experience. Science of the Total Environment 407(13) 3911-3922.

Vymazal, J., 2010b. Constructed Wetlands for Wastewater Treatment: Five Decades of Experience†. Environmental Science & Technology 45(1) 61-69.

WHO, UNICEF, 2000. Global water supply and sanitation assessment 2000 report.

WSSCC, 2004. Resource packs on the Water and Sanitation Millennium Development Goals. Water supply and sanitation collaborative council, Geneva.

Wynn, T.M., Liehr, S.K., 2001. Development of a constructed subsurface-flow wetland simulation model. Ecological Engineering 16(4) 519-536.

Chapter 2: Use of the macrophyte *Cyperus papyrus* in wastewater treatment

This chapter was presented as: Mburu, N., D.P.L. Rousseau, J.J.A van Bruggen, P.N.L. Lens (2012). Potential of *Cyperus payrus* macrophyte for wastewater treatment: A review. 12th International Conference on Wetland Systems for Water Pollution Control, Venice (Italy). 4th-9th October 2010.

Abstract

Cyperus papyrus, commonly referred to as papyrus, belongs to the Cyperaceae family and is one of the most prolific emergent macrophytes in African subtropical and tropical wetlands. Botanical studies have shown that stands of papyrus are capable of accumulating large amounts of nutrients and have a high standing biomass. Its C_4 photosynthetic pathway makes *C. papyrus* highly productive with dry weight biomass generation rates of up to 6.00 kg m^{-2} y^{-1} and nutrient uptake rates of up to 7.10 kg ha^{-1}day and 0.24 kg ha^{-1}day of, respectively, nitrogen and phosphorus. *C. papyrus* plants take about 6 - 9 months to mature with a highly reliable natural re-growth and replenishment on a site after harvesting.

Studies featuring side by side investigations with unplanted controls, show that *C. papyrus* has mostly a positive effect on the treatment of wastewater, i.e. it supports higher treatment efficiencies for the removal of organics (COD, BOD$_5$), pathogens, heavy metals and nutrients such as nitrogen and phosphorus. The ability of *C. papyrus* to use nutrients from the wastewater and the incorporation of heavy metals and organics into its phytomass, added to its easy management by regular harvesting, make it one of the most suitable plants to be used in wastewater phytoremediation in tropical areas. Therefore, it continues to be an excellent candidate for application as a macrophyte in the constructed wetland wastewater treatment technology. As such, determining the potential scope of the performance of *C. papyrus* is vital for the optimal application and design of *C. papyrus* based constructed wetland systems. This work collates growth, productivity and performance information from various independent studies incorporating the *C. papyrus* macrophyte in constructed wetlands for wastewater treatment.

2.0   Introduction

In the subtropical and tropical climate, *C. papyrus* is one of the most interesting macrophytes because it is among the most productive plants in wetlands (Kansiime et al., 2005; Heers, 2006; Perbangkhem and Polprasert, 2010). This plant has a high potential of producing biomass from solar energy, which is one of the recommended criteria for the selection of macrophytes in tropical areas with abundant sunshine for use in constructed wetlands (Perbangkhem and Polprasert, 2010). The papyrus vegetation has been shown to actively improve wastewater quality through contribution to the removal of organic compounds (Vymazal and Kröpfelová, 2009), heavy metals (Sekomo, 2012), pathogens (Kansiime and

Nalubega, 1999; Okurut, 2000) and excess nutrients such as nitrogen and phosphorus (Kansiime et al., 2007a; Kansiime et al., 2007b; Perbangkhem and Polprasert, 2010). This could be attributed to their ecological characteristics of high phytomass, well developed root system and high photosynthetic rate (Jones, 1988; Muthuri et al., 1989; Kansiime et al., 2007a). Various authors have presented experimental research findings on aspects of the characteristics of *C. papyrus* that have an influence on water quality improvement, both in the natural and constructed wetlands. This chapter collates these findings by answering the following questions related to the application and management of the macrophyte *C. papyrus* in wastewater treatment:

(a)   What is the growth habitat of *C. papyrus* macrophytes, its morphology and physical effects on water quality improvement including the surface area for attachment of microbial growth?

(b)   What is the influence of the metabolism of *C. papyrus on* water quality improvement (i.e. the plant nutrients (nitrogen and phosphorus) uptake potential, the root oxygen leakage and the biomass productivity)?

(c)   What is the harvesting practice and the regeneration capacity for *C. papyrus* after harvesting?

2.1   Influence of macrophyte on pollutant bioconversion and removal in treatment wetlands

The biogeochemical cycling and storage of nutrients, organic compounds, and metals in natural wetlands is mimicked in constructed wetlands, through the use of plants, porous media and associated microorganisms (Hunter, 2001; Sonavane, 2008). The presence of emergent macrophytes is one of the most conspicuous features of constructed wetlands and their presence distinguishes constructed wetlands from unplanted soil filters or lagoons (Greenway, 2007; Vymazal, 2011). Their positive role on the performance of constructed wetlands has been well established in numerous studies measuring treatment with and without plants (Akratos and Tsihrintzis, 2007; Yang et al., 2007; Brisson and Chazarenc, 2009; Kadlec and Wallace, 2009).

Generally, the performance of wetlands for wastewater treatment depends on the growth potential and ability of macrophytes to develop sufficient root systems for microbial attachment and material transformations, and to incorporate nutrients into plant biomass that can be subsequently harvested for nutrient removal (Kyambadde et al., 2004a; Vymazal and Kröpfelová, 2009). However, empirical exploitation of plants is a common practice. Availability, expected water quality, normal and extreme water depths, climate and latitude, maintenance requirements and project goals are among the variables that determine the selection of plant species for constructed wetlands (Stottmeister et al., 2003).

While there is a recognition that the improvement of water quality in treatment wetland applications is primarily due to microbial activity (Faulwetter et al., 2009; Kadlec and Wallace, 2009), experience has shown that wetland systems with vegetation or macrophytes has a higher efficiency of water quality improvement than those without plants (Coleman et al., 2001; Tanner, 2001; Brisson and Chazarenc, 2009). The emphasis of constructed wetland technology to date has been on soft tissue emergent plants including *Cyperus papyrus*, *Phragmites*, *Typha* and *Schoenoplectus* (Okurut, 2000; Kadlec and Wallace, 2009). These are fast growing species that have lower lignin contents and are adaptable to variable water depths. The productivity of emergent macrophytes is the highest among the aquatic plant communities in the tropics as well as in temperate regions. Emergent macrophytes are characterized by a photosynthetic aerial part above the water surface and a basal part rooted in the water substrate.

Emergent macrophytes find application in both surface and subsurface flow configurations of constructed wetlands. The significance of the plants used for wastewater purification has been emphasized by previous researchers (Brix, 1997; Peterson and Teal, 1995; Gersberg et al., 1983). Vymazal (2011) summarized the various roles played by emergent macrophytes in different configurations of constructed wetlands (Table 2.1).

Table 2.1. Major roles of macrophytes in constructed treatment wetlands (Vymazal, 2011)

Macrophyte property	Role in treatment process
Aerial plant tissue	Light attenuation—reduced growth of photosynthesis
	Influence of microclimate—insulation during winter
	Reduced wind velocity—reduced risk of re-suspension
	Aesthetic pleasing appearance of the system
	Storage of nutrients
Plant tissue in water	Filtering effect—filter out large debris
	Reduced current velocity—increased rate of sedimentation, reduced risk of resuspension
	Excretion of photosynthesis oxygen—increased aerobic degradation
	Uptake of nutrients
	Provision of surface for periphyton attachment
Roots and rhizomes in the sediment	Stabilizing the sediment surface—less erosion
	Prevention of the medium clogging in vertical flow systems
	Provision of surface for bacterial growth
	Release of oxygen increases degradation (and nitrification)
	Uptake of nutrients
	Release of antibiotics, phytometallophores and phytochelatins

Macrophyte plants, in addition to their site specific roles (i.e. attenuation of light, water current and wind velocity, aesthetic appearance, etc) are essential in the wetland treatment systems because they have properties that foster many wastewater treatment processes (Kyambadde, 2005; Kadlec and Wallace, 2009). Aquatic plants can absorb inorganic (nutrients, metals, etc), and organic pollutants (aromatics, hydrocarbons, etc) from wastewater and incorporate them into their own structure (Haberl et al., 2003), thus providing a storage and a release of nutrients through the plant growth cycle (NAS-NRC, 1976; Shetty, 2005). They create favorable conditions for microbes that contribute to the processing of pollutants by influencing the oxygen supply to the water, providing attachment surfaces, providing carbon and electron donor via carbon content of litter and root exudates (Brix and Schierup, 1989; Kadlec and Wallace, 2009). Further, aquatic plants promote stable residual accretions in the wetland (Greenway, 2007; Vymazal, 2007). These residuals contain pollutants as part of their structure or in absorbed form, and hence represent a burial process of contaminants (Kadlec and Wallace, 2009). These facts have been exploited in constructed wetland systems which have been widely used during the past decades for the treatment of wastewater because of their good efficacy to improve water quality at low operational costs (Vymazal, 1999; Neralla et al., 2000; Rousseau et al., 2004; Molle et al., 2005 ; Zurita et al., 2008; Perbangkhem and Polprasert, 2010). The natural wetlands too have been shown to have potential as sink and buffering site for organic and inorganic pollutants (Buchberger and Shaw, 1995; Muthuri and Jones, 1997; Mannino et al., 2008; Sekomo, 2010).

Wetland vegetation occasion evapotranspiration, and a corresponding increase in the hydraulic retention time that can be explained by a net biomass productivity accompanied by transpiration (Kansiime and Nalubega, 1999; Kyambadde et al., 2005). Emergent macrophyte vegetation tends to increase rates of water loss through evapotranspiration when compared to rates of evaporation from bodies of open water (Jones and Humphries, 2002).

At present there is no clear evidence that treatment performance is superior or different between the common emergent wetland species used in treatment wetlands (IWA, 2000; Zhu et al., 2010). Even so some soft tissue emergent macrophytes including *Phragmites* sp., *Schoenoplectus* sp., *Typha* sp. and *Carex* (true sedge) are well known for their potentials in constructed wetlands treating wastewater and fecal sludge, and their performances are well documented, especially for the high latitudes, temperate climate regions (Fennessy et al., 1994; Coleman et al., 2001; Ciria et al., 2005; Stein et al., 2006). These macrophytes are, however, not found in all regions of the world and efforts are being made worldwide to select candidate macrophytes to be exploited locally in constructed wetlands (Azza, 2000; Yang et al., 2007; Brisson and Chazarenc, 2009; Huang, 2010; Perbangkhem and Polprasert, 2010)

2.2   *Cyperus papyrus* macrophyte

2.2.1   History and growth habitat
C. papyrus, commonly called papyrus or paper plant, is a member of the *Cyperaceae* sedge family, a group of plants closely related to the grasses (Michael, 1983). The *Cyperaceae* family has about 75 genera and more than 4000 species, which are for a large part perenial rhizomateous herbs growing in moist places. *C. papyrus* has a long history of being harvested and has been used over millennia, such as for the manufacture of the first paper by the ancient Egyptians (Terer et al., 2012). It once grew wild throughout the Nile Valley (Egypt, Ethiopia), and can still be found in the swamps and marshes of Central, East and Southern Africa (Chale, 1987; Boar et al., 1999; El-Ghani et al., 2011; van Dam et al., 2011). It was widely cultivated in Egypt for its many uses: boats, rope, food (boiled pith and rhizomes), sandals, boxes, mats, sails, blankets, cloth, mummy wrappings, firewood (dried rhizomes), medicine, and building materials as well as writing materials (scrolls *Papyri*) (Leach and Tait 2000). It is the largest of the sedges, and a monocot that is native to riverbanks and mouths, lakeshores, floodplains and wet soil areas of North and tropical Africa. Outside Africa, it is thought to be native to the Hula Valley in Israel where it reaches its northernmost limits. It

has been introduced and naturalised in the Mediterranean (Sicily, Malta), USA (Florida) and India (Terer et al., 2012).

C. papyrus is the dominant species of many swamps in East and Central Africa and can be found growing in both lentic and lotic fresh water environments with stable hydrological regimes (permanently flooded). It cannot cope with rapid water level changes and water flow (Kresovich et al., 1982; Jones and Muthuri, 1985; Jones, 1988; Serag, 2003). Due to its rhizomatic root structure it can also be found floating with a mat-like root structure (up to 1.5 m thick) in open waters as deep as 3-4 m (Thompson, 1979; Kansiime and Nalubega, 1999; Azza, 2000). As a member of the sedge family it does not hold economic importance as a crop plant, nevertheless in some regions it still finds application in weaving mats, baskets, screens, and even sandals (Osumba et al., 2010; van Dam et al., 2011; Morrison et al., 2012).

A substantial number of sedges are weeds, invading crop fields in all climates of the world. Sedges do, however, have a considerable ecological importance. They are of extreme importance to primary production as well as an integral part of the hydrologic cycle (Saunders et al., 2007). Today, the most important uses of papyrus wetlands are those of ecological resources and services (Maclean et al., 2011; van Dam et al., 2011).

The C. papyrus wetland soils and plants may absorb or adsorb heavy metals, pathogens, inorganic forms of nitrogen, phosphorus, other nutrients and trace elements. The rhizomes of the plant prevent soil erosion, and trap polluted sediments from inflowing water. Consequently, C. papyrus has found application in both constructed and natural wetlands for water quality improvements. Thus, even in modern times papyrus may have an important role in cleaning up wastewater pollution from industrial, municipal and domestic sources as captured in the listed selection of studies in Table 2.2. Side by side investigations with unplanted controls shows the macrophyte C. papyrus has mostly a positive effect, i.e. supports higher treatment efficiency for the removal of organics (COD, BOD_5), faecal coliforms, heavy metals and nutrients such as nitrogen and phosphorus (Nyakang'o and van Bruggen, 1999; Okurut, 2000; Kyambadde et al., 2005; Abira, 2007; Sekomo, 2012).

In constructed wetland applications, C. papyrus has been found to establish well from rhizome fragment propagules and also to adapt well to wastewater conditions (Okurut, 2000; Abira, 2007 Mburu et al., 2013). This characteristic of vegetative reproduction via rhizomes and rapid recovery after damage to aboveground growth is shared by other effective invasive macrophytes such as *Phragmites* (Meyerson et al., 2000).

2.2.2 *Cyperus papyrus* morphology

C. papyrus has its culm or stem (often triangular) growing to an average height of 3-5 m above ground and taking 6-9 months to mature (Gaudet, 1977; Kresovich et al., 1982; Muthuri and Jones, 1997; Terer et al., 2012). The culm has a large proportion of a spongy aerenchyma on its inside and to a small extent, it is capable of photosynthesis (Okurut, 2000). It is topped by characteristically large, spherical shaped (finely dissected bracteoles), reproductive umbels (it bears flowers) that serve also as main photosynthetic surface of the plant (Jones and Humphries, 2002; Mnaya et al., 2007). The rhizomes and the roots together form a mat like structure that is the base for swamp development. *C. papyrus* can grow well in the subtropical and tropical climate and is among the most productive plants of wetlands (Kansiime et al., 2005; Heers, 2006; Perbangkhem and Polprasert, 2010).

C. papyrus is considered to be unique due to its C_4 photosynthetic pathway in spite of the fact that it grows in a wetland ecosystem, which appears an unlikely habitat for C_4 species (Jones, 1987; 1988; Jones and Humphries, 2002; Saunders et al., 2007). Plants utilizing the C_4 photosynthetic pathway show higher potential efficiencies in the use of intercepted radiation, water, and nitrogen for the production of dry matter than do other photosynthetic types (Piedade et al., 1991). C_4 species are most numerous in tropical and warm, temperate semi-arid zones, where their greater water-use efficiency appears to be a key selective advantage (Piedade, 1991). The C_4 photosynthetic pathway makes *C. papyrus* highly productive with dry weight biomass generation of up to 6.28 kg m^{-2} y^{-1} (Terer et al., 2012).

Aerobic conditions in the roots and rhizomes of *C. papyrus* are maintained by oxygen transport from the atmosphere through the aerenchyma of the culms (Li and Jones, 1995).

Table 2.2. Application and investigations of *Cyperus papyrus* macrophyte for water quality improvement

Location	Set-up	Area	Type of wastewater	Removal efficiency	Reference
Kenya	Experimental CW cells	3.2x1.2x0.8 m	Pulp and paper mill	*TN(75%),*BOD-90%,*TSS-81%;	Abira (2007)
Uganda	Experimental CW cells	60m²	Urban sewage	*COD-52%; Phenols-73-96% BOD,NH₄⁺-N,P: 68.6-86.5%	Kyambadde et al. (2005)
Kenya (Kahawa swamp)	Natural wetland	3.7ha	Domestic wastewater	DO, NH₄⁺-N,ortho-P:77-85%	Chale(1985)
Uganda (Nakivubo swamp)	Natural wetland	2.5km²	Urban sewage	*NH₄⁺-N (89.4)*ortho-P (80) *COD (70%);	Kansiime et al. (2003)
Uganda	Experimental pilot-scale CW	320m²	Domestic wastewater	TSS(80%);*Coliforms(4log units)	Okurut (2000)
Congo (Upemba wetlands)	Rectangular transects, 2 x 10-20 m	250m²	Natural wetlands		Thomson et al. (1979)
Kenya	Operation CW	0.5 ha	Domestic wastewater		Nyakang'o et al.1999
Thailand	Experimental pilot-scale CW	3m²	Domestic wastewater		Perbangkhem et al. (2009)
Kenya (Lake Naivasha)	Quadrants of 3 m x 3 m (lake-)		Natural wetland		Muthuri et al (1989)
Tanzania	0.5 m X 0.25 m quadrants	1.41 km²	Natural wetland (Rubondo Island)		Mnaya et al. (2007)
Kenya	0.5 m X 0.5 m quadrants	150km²	Natural wetland (Lake Naivasha)		Boar (2006)
Kenya (shores of Lake Victoria)	2 m x 2 m quadrants	60 km stretch	Natural wetland		Osumba (2010)
Rwanda (Nyabugogo wetland)	13 sampling sites	60 ha	Municipal/Industrial (heavy metals)	4.2 mg/kg for Cd, 45.8 mg/kg for Cr, 29.7 mg/kg for Cu and 56.1 mg/kg for Pb	Sekemo et al.(2010)
Uganda (Natete wetland)	Transects (325m, 350 m long) 50 sites (geo referenced using a GPS)	1 km²	Municipal	NH₄⁺-N (21%), NO₃-N (98%), TN (35%)	Kanyiginya et al.(2010)
Nile Delta/ Egypt	Yard-scale experiment	2250km²	Natural wetland		El-Ghani et al.(2011)
Cameroon		1m²	Fecal sludge		Kengne et al. (2008)
USA	Microcosms	0.49 x 0.35 x	Synthetic domestic wastewater		Morgan et. al. (2008)
China	Microcosms	0.26 m²	Artificial wastewater		Wang et al.(2008)

*: Maximum values

Aerenchymous plant tissue is an important adaptation to flooding in wetland plants through which transport of gases to and from the roots through the vascular tissues of the plant above water and in contact with the atmosphere takes place (Singer et al., 1994). Li and Jones (1995) reported a diffusive oxygen transport between the rhizomes and the culms of 2.45 and 3.29 mol m^{-3} of oxygen at day and night time, respectively. This provides an aerated root zone and thus lowering the plant's reliance on external oxygen diffusion through water and soil (Kadlec and Wallace, 2009).

In the tropical swamps *C. papyrus* establishment, growth and mortality occur concurrently through the year, so that there is little temporal change in the standing crop (Muthuri et al., 1989). The culms can be divided into different age classes; some authors have used classification based on three age classes, namely juvenile, with unopened umbels, mature, with opened green umbels and senescent with more than half of the umbels brown (achlorophyllous) (Muthuri et al., 1989), while others have identified six age classes namely, young elongated culm with closed umbel, elongated culm with umbel just opening, fully elongated culm and fully expanded umbel, fully elongated culm and fully expanded umbel but older, senescing culm (> 10% achlorophyllous), dead culm (> 80% achlorophylious) (Muthuri and Jones, 1997). Culm density is controlled by density dependent mortality (Thompson, 1979).

2.3 *Cyperus papyrus* biomass productivity

In both natural and constructed wetlands with *C. papyrus,* vegetative and reproductive parts above the ground level and their root systems comprise a substantial part of the wetland biomass (Fig.2.1). Emergent macrophytes in swamps and marshes are amongst the most productive plant communities (Muthuri et al., 1989). *C. papyrus* vegetation is highly productive and under favourable temperatures, hydrological regime and nutrient availability estimates of aerial standing live biomass (including scale leaves, culm and umbel) often exceed 5000 g (dry weight) m^{-2} (Saunders et al., 2007).

Fig. 2.1. Photos of the above ground vegetative and reproductive parts of *Cyperus papyrus* in a constructed wetland at Juja, Kenya

The productivity of natural papyrus wetlands is found to be variable (Table 2.3) and controlled by different factors such as climate, nutrient availability and the prevailing general hydrological conditions (Okurut, 2000). Differences in aerial biomass of papyrus in various sites have been attributed to prevailing climatic conditions. Some studies have noted a trend of an increase in standing biomass of papyrus swamps with increase in altitude. Nevertheless the trend has not been found to hold by all authors (Thompson, 1979; Muthuri et al., 1989; Mnaya et al., 2007). Unlike other emergent aquatic plants (Table 2.4), its high productivity rates and standing / harvestable biomass makes *C. papyrus* have a high nutrient removal potential more so in wetlands receiving a high nutrient load. The harvesting of biomass presents a potential for biological nutrient removal (Kansiime and Nalubega, 1999; Kyambadde et al., 2005).

Estimating biomass or primary productivity in tropical swamps which have relatively stable biomass, requires measurements of population dynamics and the life cycle of individual shoots (Muthuri et al. 1988), unlike in the temperate ecosystems, where common methods of estimating primary productivity include measurements of peak biomass, maximum minus minimum biomass or methods which account for death and decomposition between harvests (Muthuri et al., 1989; Sala, 2000).

2.2.3.1   Above ground biomass

The aerial organs of *papyrus* (umbel, culm and scale leaves) contribute about 50 % of the total plant biomass (Thompson, 1979), in which the largest proportion is in culms (Muthuri et al., 1989). The high aerial biomass of papyrus (Table 2.3) is unlike many other perennial emergent macrophytes such as *Typha latifolia* (890-2500 g m^{-2}), *Scirpus validus* (2355-2650 g m^{-2}) and *Phragmites australis* (1110-5500 g m^{-2}) (Kadlec and Wallace, 2009) that have a large proportion of their biomass in the form of roots and rhizomes (Muthuri et al. 1988). High aerial primary production indicates that less carbohydrate is assimilated in the rhizomes, hence the living culms act as storage organs. This function is normally for the rhizome (Tanner, 1996). Boar *et al.* (1999) established that biomass allocation to the various *C. papyrus* tissues is directly related to the fertility of the growing media such that least investment in roots and rhizome indicates plenty of nutrient supply. However, some studies have found that an important fraction of the plants' biomass is stored in the below-ground stands of the papyrus (Saunders et al., 2007; Kengne et al., 2008; Kanyiginya et al., 2010).

2.2.3.2 Below ground biomass

The below ground biomass of *C. papyrus* consists of an interlaced but permeable root mat with a rhizomatic structure (Kansiime and Nalubega, 1999). Measurements of rhizomes and root mass in the papyrus vegetation involve excavation to the maximum depth to which the roots are found (Muthuri et al., 1989). In natural swamps, the rooting mat has been estimated to contribute up to   30 - 52 % of the total biomass (Boar et al., 1999; Okurut, 2000). The below-ground biomass (i.e. the root and rhizomes) surface area provides attachment sites which are conducive for the proliferation of bacterial biomass. The roots and rhizomes influence the wastewater residence time, trapping and settling of suspended particles, surface area for pollutant adsorption, uptake, assimilation in plant tissues and oxygen for organic and inorganic matter oxidation in the rhizosphere (Kansiime and Nalubega, 1999; Okurut, 2000; Kyambadde et al., 2004a). For example, the nature and density of the rooting biomass can greatly influence the extent of faecal bacteria removal via sedimentation and attachment processes. This influence was demonstrated in the studies of Kansiime and Nalubega (1999) in a natural wetland where faecal coliform counts were consistently higher in zones dominated by the *Miscanthidium violaceum* macrophyte, than in zones dominated by *C. papyrus*. The rooting mat of the former was tight and compact and thus had a reduced total surface area. In contrast, the papyrus mat is hollow and interwoven giving it a larger surface area for entrapment and attachment of faecal coliforms (Okurut, 2000). Sekomo (2012) established that *C. papyrus* plants plays an

24

important role in metal retention. The *C. papyrus* root system was the most important part of the plant in heavy metal retention, followed by the umbel and finally the stem.

Table 2.3. Biomass productivity of *Cyperus papyrus* growing in different types of wetlands

Study/ Site	Biomass productivity (Dry weight g biomass m^{-2})		Reference
	Below ground	Aerial	
Pilot scale (Free water Surface), Uganda	1250	2250	Kyambadde et al. (2005)
Natete wetland, Kampala, Uganda	1288 ± 8	1020±14	Kanyiginya et al. (2005)
Man made swamp, Kenya (Kahawa Swamp)	4,955*		Chale (1987)
Lake Naivasha, Kenya		3245	Jones and Muthuri (1985)
Busoro (Flooded river valley), Rwanda		1384	
Nakivubo wetland (two sites), uganda		883-1,156 3,529-5,844	Kansiime et al. (2003)
Pilot Scale Constructed Wetland, Uganda			Nyakang'o et al.1999
Constructed wetland, Thailand	2200–3100*		Perbangkhem et al. (2009)
Constructed wetlands, Uganda	16900-18700*		Kansiime et al. (2005)
Natural wetland, Lake Naivasha, Kenya		2731	Muthuri et al.(1989)
Natural wetland, Lake Naivasha, Kenya		4652	Terer et al. (2012)
Rubundo Island,Lake Victoria, Tanzania	4144 ± 452	5789 ± 435	Mnaya et al. (2007)
Lake Naivasha, Kenya	6950 ± 860*		Boar (2006)
Nakivubo wetlands, Uganda	6700*		Kansiime et al. (2007)
Kirinya Wetlands, Uganda	7200*		
Nakivubo wetland, Uganda	1158	2480	Mugisha et al. (2007)
Kirinya wetland, Uganda	4343	3290	Mugisha et al. (2007))

*Total biomass (Below ground + Aerial)

Table 2.4. Biomass productivity of other macrophytes growing in different types of wetlands

Macrophyte	Study/ site	Biomass productivity (Dry weight g biomass m^{-2})		Reference
		Below ground	Aerial	
Colocasia esculenta	Nakivubo wetland, Uganda	1236	2024	Mugisha et al. (2007)
Colocasia esculenta	Kirinya wetland, Uganda	1697	2463	Mugisha et al. (2007)
Miscunthus Violeceus	Nakivubo wetland, Uganda	870	1190	Mugisha et al. (2007)
Miscunthus Violeceus	Kirinya wetland, Uganda	1470	1680	Mugisha et al. (2007)
Phragmites mauritianus	Nakivubo wetland, Uganda	745	1790	Mugisha et al. (2007)
Phragmites mauritianus	Kirinya wetland, Uganda	1452	3030	Mugisha et al. (2007)
Phragmites australis	Tidal salt marsh, North America		727-3663	Meyerson et al. (2000)
Phragmites australis	Freshwater marsh, N. America		980-2642	Meyerson et al. (2000)

2.2.4 Nutrient uptake and storage

The removal of soluble inorganic nitrogen and phosphorus via absorption from either the water column or the sediment and storage in plant tissue is a direct mechanism of nutrient sequestration (Greenway, 2007). Table 2.5 shows values for nutrient uptake of *C. papyrus* under different set-ups. The difference in the uptake rates may be attributed to nutrient availability under the experimental conditions and/ or the growth phase of the macrophyte. Comparison of nutrient concentrations in plants, soil and water column in the Natete wetland (Kampala, Uganda), found that *C. papyrus* stored the highest amounts of nutrients as compared to soil and water (Kanyiginya et al.,

2010). Plants take up nutrients as a requirement for their growth. These nutrients accumulate in plant parts which present an opportunity to remove excess nutrients from wetland systems through harvesting the aerial plant phytomass (Kansiime et al., 2007a). In this regard, plants with high rates of net primary productivity and higher nutrient uptake are preferred in wetlands subject to wastewater inputs (Kansiime et al., 2007).

Table 2.5. *Cyperus papyrus* macrophytes nutrient uptake rates under varying experimental set-ups

Type of Wastewater	Phosphorus uptake (Kg ha^{-1} day^{-1})	Nitrogen uptake (Kg ha^{-1} day^{-1})	Reference
Septic tank effluent (Constructed wetland, Uganda)	0.24	7.1	Okurut (2000)
Natural wetland (Lake Naivasha)	0.06	1.18	Muthuri et al. (1989)
Municipal sewage (Nakivubo wetland)	0.21	1.3	Kansiime et al. (2003)
Natural wetland (Upemba swamps)	0.06	1.18	Thompson et al. (1979)
Domestic wastewater (Constructed Wetlands)	0.14	3.01	Brix (1994)
Other macrophytes			
Phragmites australis (Infiltration wetland)	0.22	2.14	Okurut (2000)
Eichhomia crassipes (diverse wastewater)	0.2-2	1.6-6.6	Okurut (2000)

The nutrient elements essential for plant growth would be removed in proportion to their compositional ratios in the particular species (Boyd, 1970). For *C. papyrus,* Chale (1987) found the nitrogen concentrations of the various plant organs were 4.8% roots, 8.4% rhizomes, 4.5% scales, 4.8% culms, and 6.2% umbels on dry weight basis. As to phosphorus, the concentrations were 0.09% roots, 0.11% rhizomes, 0.09% scales, 0.10% culms, and 0.13% umbels. A high content of nutrients is observed for the aerial biomass of papyrus, an indication of active translociion and storage of nutrients to parts of the plant where they are needed for primary growth, e.g. synthesis of amine acids and enzymes (Muthuri and Jones, 1997; Kyambadde et al., 2005; Kansiime et al., 2007b). Significantly higher amounts of nutrients are sequestered in papyrus umbels and culms compared to roots/rhizomes portions (Kyambadde et. al., 2005). Similar observations have been made by Mugisha *et al.* (2007) who established that photosynthetic organs of *C. papyrus* (culm and umbel) generally had a higher nutrient content than other organs (roots and rhizome) at the Nakivubo and Kirinya wetlands at the shores of Lake Victoria in Uganda. Nevertheless, nutrients in papyrus plants decrease with the age of the plant as the nutrients are translocated to the metabolically active juvenile plants for growth (Mugisha et al., 2007).

Okurut (2000) found nutrient removal from wastewater via plant uptake to show variability at different growth phases. The growth rate of *Cyperus papyrus* is the highest in juvenile plants and the lowest in mature plants and the nitrogen uptake rate by *Cyperus papyrus* is the highest in juvenile plants and the lowest in mature plants. Uptake was correlated with the biomass yields exhibited in the different phases. The total nitrogen content was the highest in the juvenile plants and decreased with increasing age. This enables the plant to recycle nutrients from the old portions to new growth (Boyd, 1970). Generally, (a) the rate of nutrient uptake by macrophytes is limited by its growth rate and the concentration of nutrients within the plant tissue and (b) nutrient storage is dependent on the plant-tissue nutrient concentration and potential for biomass accumulation (Greenway, 2007).

2.3   Wastewater treatment with *Cyperus papyrus*

C. papyrus plays an important role in the water quality enhancement, the effects of which can be readily observed in terms of dissolved oxygen (DO), pH and redox potential of their surroundings and the attenuation of pollutant parameter profiles from influent to effluent (Okurut, 2000; Huang, 2010). For constructed wetlands to be effective in water pollution control, they must function as "pollutant" sink for organics, sediments, nutrients and metals, i.e. these pollutants must be transformed, degraded or removed from the wastewater and stored within the wetland either in the sediment or the plants. Although there is still debate about the relative importance of macrophytes versus microbes in nutrient removal, plant biomass still accounts for substantial removal and storage of nitrogen and phosphorus (Brix, 1997; IWA, 2000).

Macrophytes can contribute directly through uptake (nutrients and heavy metals), sedimentation, adsorption or phytovolatilization or indirectly to pollutant removal in constructed wetlands. Indirect processes are related to biofilm growth around roots, evapotranspiration, and the pumping of oxygen towards the rhizosphere that changes the redox conditions (Imfeld et al., 2009; Kadlec and Wallace, 2009; Carvalho et al., 2012). Some of these mechanisms are addressed in the sections below.

2.3.1 Root oxygen release into the rhizosphere

Papyrus-dominated wetlands like all other natural wetlands are characterized by low dissolved oxygen concentrations (Okurut, 2000). The main reason for this state is that surface aeration and photosynthetic oxygen transfer mechanisms are poor or non existent due to the dense plant canopy. On the other hand oxygen leakage to the rhizosphere is important in constructed wetlands with subsurface flow for aerobic degradation of oxygen-consuming substances and nitrification (Brix, 1994). The photosynthetic characteristics of wetland species can affect their ability to provide oxygen, and this ultimately influences their disposal efficiencies.

The peak photosynthetic quantum efficiency, i.e. the amount of CO_2 that is fixed or the amount of O_2 that is released via assimilation when the photosynthetic apparatus absorbs one photon (Huang, 2010) for *C. papyrus* has been reported to range between 26 and 40 µmol CO_2 m^{-2} s^{-1} (Jones, 1987; 1988; Saunders and Kalff, 2001). In their work on plant photosynthesis and its influence on removal efficiencies in constructed wetlands, Huang et al. (2010) published the photosynthetic rates of five wetland plants, *Phragmites, Ipomoea, Canna, Camellia,* and *Dracaena,* at light saturation. These ranged between 11.6 and 31.32 µmol CO_2 m^{-2} s^{-1}. *C. papyrus* presents a comparable potential for oxygen production via photosynthesis. Kansiime and Nalubega (1999) estimated oxygen release rates of 0.017 g m^{-2} day^{-1} by C. *papyrus* plants. The oxygen released is available for microbiota within the rhizosphere.

2.3.2 Surface for microorganism's attachment

In natural and constructed wetlands, macrophyte root structures provide microbial attachment sites. In an experimental microcosm set-up, Gagnon et al. (2007) found that microbes were present on substrates and roots as an attached biofilm and abundance was correlated to root surface throughout depth. Indeed planted wastewater treatment systems outperform unplanted ones, mainly because plants stimulate below ground microbial populations (Gagnon et al., 2007). Plant species root morphology and development seem to be a key factor influencing microbial– plant interactions. Kyambadde et al. (2004) measured a higher root surface and microbial density in a constructed wetland planted with *C. papyrus* (average root surface area 208.6 cm^2) compared to *Miscanthidium violaceum* (average root surface area 72.2 cm^2). *C. papyrus* and *Miscanthidium violaceum* differed in their root recruitment rate and root number in a microcosm constructed wetland. The root recruitment rate per constructed wetland unit was 77 and 32 roots per week for *C. papyrus* and *Miscanthidium violaceum*, respectively, and *C. papyrus* had more adventitious roots and larger root surface area than *Miscanthidium violaceum* (Kyambadde et al., 2004b). Further, *C. papyrus* seems to promote greater nitrogen removal efficiencies, through nitrification and denitrification rates of bacteria associated with it roots (Morgan et al., 2008).

2.3.3   Evapotranspiration

The average daily water vapour flux from the papyrus vegetation through canopy evapotranspiration in a wetland located near Jinja (Uganda) on the Northern shore of Lake Victoria was approximated by Saunders et al. (2007) as 4.75 kg H$_2$O m^{-2} d^{-1} (= 4.75 mm/d), which was approximately 25% higher than water loss through evaporation from open water (approximated as 3.6 kg H$_2$O m^2 d^{-1}). Jones and Muthuri (1985) reported an evapotranspiration rate of 12.5 mm/day at the fringing papyrus swamp on Lake Naivasha, while Kyambadde et al. (2005) reported 24.5±0.6 mm/d for a subsurface horizontal flow wetland in Kampala (Uganda). Evapotranspiration rates vary sharply since they depend on numerous factors influencing the ecosystem's prevailing micro-climate, as listed by Kadlec and Wallace (2009). For example, common reed transpiration rates oscillate between 4.7-12.4 mm/day depending on meteorological conditions (Holcová et al., 2009). Evapotranspiration (ET) by plants can significantly affect the hydrological balance of treatment wetlands. The water lost through ET concentrates pollutants within the wetland, while the volume reduction results in longer hydraulic retention times (Kadlec and Wallace, 2009). For low-loaded systems or systems with longer retention times, such evapotranspiration rates can exceed the influent wastewater flow, leading to a zero discharge.

2.4   *Cyperus papyrus* **harvesting and regeneration potential**

In order to achieve a permanent nutrient removal from wetland systems, *C. papyrus* harvesting is encouraged, but this requires careful timing (Kiwango and Wolanski, 2008). Total nitrogen in aerial biomass of *C. papyrus* decreases from the juvenile plants to older plants (Mugisha et al., 2007). Hence, to minimize internal nutrient cycling and eventual export of the nutrients from wetland systems, sustainable harvesting of aerial biomass at different growth stages can be used as a strategy to reduce nutrients, especially in wastewater treatment wetlands. The regeneration potential of *C. papyrus,* i.e. the inherent capacity for natural re-growth and replenishment, on a site after a disturbance has been found to be highly reliable (Osumba et al., 2010). However, overharvesting (within less than one 6-months growing season) of *C. papyrus* can reduce this regeneration potential leading to weak spatial connectivity, papyrus stand fragmentation and increased landscape patchiness in natural wetlands (Osumba et al., 2010). Modeling studies of papyrus wetlands by van Dam *et al.* (2007) have proposed a harvesting rate between 10% and 30% of the total biomass per year. At higher harvesting rates, nutrient uptake and retention by papyrus does not increase proportionally because of reduction in plant biomass, leading to lower uptake (van Dam et al., 2007). Muthuri et al. (1988) established a ceiling aerial biomass of 2,731 g m^{-2} after six moths, at a previously harvested section of a swamp at Lake Naivasha (Kenya), while for the undisturbed sections of the swamp an aerial biomass of 3602 g m^{-2} was recorded. Water levels after harvesting are thought to affect biomass yield. Osumba et al. (2010) found that flooded sites give the least regenerated biomass yields.

2.5   Conclusion

This literature survey reveals that the macrophyte *C. papyrus* has found application in constructed wetlands for remediating a variety of pollutants in wastewater from different sources. The majority of the application of the *C. papyrus* macrophyte in constructed wetlands is found in the developing tropical countries, where papyrus is occurring locally. The macrophyte possesses a robust morphology and metabolism, it is easy to establish and manage, thus making constructed wetlands incorporating *C. papyrus* wetland vegetation a promising wastewater treatment option for wider application. The production and harvesting of vegetation biomass from these treatment wetlands can provide a permanent route for the removal of nutrients, with economic benefits for communities that engage in the trade of papyrus products.

2.6   References

Abira, M.A., 2007. A pilot constructed treatment wetland for pulp and paper mill wastewater: performance, processes and implications for the Nzoia river, Kenya. PhD Thesis, UNESCO-IHE, Delft, Netherlands.

Akratos, C.S. and V.A. Tsihrintzis, 2007. Effect of temperature, HRT, vegetation and porous media on removal efficiency of pilot-scale horizontal subsurface flow constructed wetlands. Ecological Engineering, 29: 173-191.

Azza, N., G., T., Kansiime, F., Nalubega, M., Denny, P., 2000. Differential permeability of papyrus and *Miscanthidium* root mats in Nakivubo swamp, Uganda. Aquatic Botany, 67: 169-178.

Boar, R.R., D.M. Harper and C.S. Adams, 1999. Biomass allocation in *Cyperus papyrus* in a tropical wetland, Lake Naivasha, Kenya. Biotropica, 31: 411-421.

Boyd, 1970. Vascular aquatic plants for mineral nutrient removal from polluted waters. Economic Botany, 24: 95-103.

Brisson, J. and F. Chazarenc, 2009. Maximizing pollutant removal in constructed wetlands: Should we pay more attention to macrophyte species selection? Science of the Total Environment, 407: 3923-3930.

Brix, 1997. Do macrophytes play a role in constructed treatment wetlands? Water Science and Technology, 35: 11-17.

Brix and Schierup, 1989. The use of aquatic macrophytes in water pollution control. Ambio 18: 100-107.

Brix, H., 1994. Functions of macrophytes in constructed wetlands. Water Science and Technology, 29: 71-78.

Buchberger, S.G. and G.B. Shaw, 1995. An approach toward rational design of constructed wetlands for wastewater treatment. Ecological Engineering, 4: 249-275.

Carvalho, P.N., M.C.P. Basto and C.M.R. Almeida, 2012. Potential of Phragmites australis for the removal of veterinary pharmaceuticals from aquatic media. Bioresource Technology, http://dx.doi.org/10.1016/j.biortech.2012.03.066.

Chale, F., 1987. Plant biomass and nutrient levels of a tropical macrophyte *Cyperus papyrus* receiving domestic wastewater. Aquatic Ecology, 21: 167-170.

Ciria, M.P., M.L. Solano and P. Soriano, 2005. Role of Macrophyte Typha latifolia in a Constructed Wetland for Wastewater Treatment and Assessment of Its Potential as a Biomass Fuel. Biosystems Engineering, 92: 535-544.

Coleman, J., K. Hench, K. Garbutt, A. Sexstone, G. Bissonnette and J. Skousen, 2001. Treatment of Domestic Wastewater by Three Plant Species in Constructed Wetlands. Water, Air, & Soil Pollution, 128: 283-295.

El-Ghani, M.A., A.M. El-Fiky, A. Soliman and A. Khattab, 2011. Environmental relationships of aquatic vegetation in the fresh water ecosystem of the Nile Delta, Egypt. African Journal of Ecology, 49: 103-118.

Faulwetter, J.L., V. Gagnon, C. Sundberg, F. Chazarenc, M.D. Burr, J. Brisson, A.K. Camper and O.R. Stein, 2009. Microbial processes influencing performance of treatment wetlands: A review. Ecological Engineering, 35: 987-1004.

Fennessy, M.S., J.K. Cronk and W.J. Mitsch, 1994. Macrophyte productivity and community development in created freshwater wetlands under experimental hydrological conditions. Ecological Engineering, 3: 469-484.

Gagnon, V., F. Chazarenc, Y. Comeau and J. Brisson, 2007. Influence of macrophyte species on microbial density and activity in constructed wetlands. Water Science and Technology, 56: 249-254.

Gaudet, J.J., 1977. Uptake, Accumulation, and Loss of Nutrients by Papyrus in Tropical Swamps. Ecology, 58: 415-422.

Greenway, M., 2007. The Role of Macrophytes in Nutrient Removal using Constructed Wetlands Environmental Bioremediation Technologies. In: S. Singh and R. Tripathi (Eds), Springer Berlin Heidelberg, pp. 331-351.

Haberl, R., S. Grego, G. Langergraber, R.H. Kadlec, A.R. Cicalini, S.M. Dias, J.M. Novais, S. Aubert, A. Gerth, H. Thomas and A. Hebner, 2003. Constructed wetlands for the treatment of organic pollutants. Journal of Soils and Sediments, 3: 109-124.

Heers, M., 2006. Constructed wetlands under different geographic conditions: Evaluation of the suitability and criteria for the choice of plants including productive species.Master Thesis. Hamburg University of Applied Sciences, Germany Faculty of Life Sciences.Department of Environmental Engineering

Holcová, V., J. Šíma, K. Edwards, E. Semančíková, J. Dušek and H. Šantrůčková, 2009. The effect of macrophytes on retention times in a constructed wetland for wastewater treatment. International Journal of Sustainable Development & World Ecology, 16: 362-367.

Huang, J., Wang, Shi-he, Yan, Lu Zhong, Qiu-shuang, 2010. Plant photosynthesis and its influence on removal efficiencies in constructed wetlands. Ecological Engineering, 36: 1037-1043.

Hunter, R.G., Combs, D. L.,George, D. B., 2001. Nitrogen, phosphorous, and organic carbon removal in simulated wetland treatment systems. Archives of Environmental Contamination and Toxicology, 41: 274-281.

Imfeld, G., M. Braeckevelt, P. Kuschk and H.H. Richnow, 2009. Monitoring and assessing processes of organic chemicals removal in constructed wetlands. Chemosphere, 74: 349-362.

IWA, 2000. Constructed wetlands for pollution control. Processes, performance, design and operation. Scientific and Technical report No.8.

Jones, 1987. The photosynthetic characteristics of papyrus in a tropical swamp. Oecologia, 71: 355-359.

Jones, 1988. Photosynthetic responses of C3 and C4 wetland species in a tropical swamp. Ecology 76: 253-262.

Jones and S.W. Humphries, 2002. Impacts of the C4 sedge *Cyperus papyrus L.* on carbon and water fluxes in an African wetland. Hydrobiologia, 488: 107-113.

Jones and F.M. Muthuri, 1985. The Canopy Structure and Microclimate of Papyrus (*Cyperus Papyrus*) Swamps. Journal of Ecology, 73: 481-491.

Kadlec and S. Wallace, 2009. Treatment wetlands. 2nd ed. Boca Raton, Fla: CRC Press, 1048 pp.

Kansiime, F., E. Kateyo, H. Oryem-Origa and P. Mucunguzi, 2007a. Nutrient status and retention in pristine and disturbed wetlands in Uganda: management implications. Wetlands Ecology and Management, 15: 453-467.

Kansiime, F. and M. Nalubega, 1999. Wastewater treatment by a natural wetland: the Nakivubo swamp Uganda - processes and implications. PhD Thesis, UNESCO-IHE, Delft, Netherlands.

Kansiime, F., H. Oryem-Origa and S. Rukwago, 2005. Comparative assessment of the value of papyrus and cocoyams for the restoration of the Nakivubo wetland in Kampala, Uganda. Physics and Chemistry of the Earth, Parts A/B/C, 30: 698-705.

Kansiime, F., M. Saunders and S. Loiselle, 2007b. Functioning and dynamics of wetland vegetation of Lake Victoria: an overview. Wetlands Ecology and Management, 15: 443-451.

Kanyiginya, V., F. Kansiime, R. Kimwaga and D.A. Mashauri, 2010. Assessment of nutrient retention by Natete wetland Kampala, Uganda. Physics and Chemistry of the Earth, Parts A/B/C, 35: 657-664.

Kengne, I.M., A. Akoa, E.K. Soh, V. Tsama, M.M. Ngoutane, P.H. Dodane and D. Koné, 2008. Effects of faecal sludge application on growth characteristics and chemical composition of *Echinochloa pyramidalis* (Lam.) Hitch. and Chase and *Cyperus papyrus* L. Ecological Engineering, 34: 233-242.

Kiwango, Y.A. and E. Wolanski, 2008. Papyrus wetlands, nutrients balance, fisheries collapse, food security, and Lake Victoria level decline in 2000-2006. Wetlands Ecology and Management, 16: 89-96.

Kresovich, S., C.K. Wagner, D.A. Scantland, S.S. Groet and W.T. Lawhon, 1982. The Utilization of Emergent Aquatic Plants for Biomass Energy Systems Development. Solar Energy Research Institute Task No. 3337.01 WPA No. 274-81.

Kyambadde, 2005. Optimizing processes for biological nitrogen removal in Nakivubo wetland, Uganda.

Kyambadde, K. J., G. F. and D. L., G., 2004a. A comparative study of *Cyperus papyrus* and *Miscanthidium violaceum*-based constructed wetlands for wastewater treatment in a tropical climate. Water Research, 38: 475-485.

Kyambadde, F. Kansiime and G. Dalhammar, 2005. Nitrogen and phosphorus removal in substrate-free pilot constructed wetlands with horizontal surface flow in Uganda. Water Air and Soil Pollution, 165: 37-59.

Kyambadde, J., F. Kansiime, L. Gumaelius and G. Dalhammar, 2004b. A comparative study of Cyperus papyrus and Miscanthidium violaceum-based constructed wetlands for wastewater treatment in a tropical climate. Water Research, 38: 475-485.

Li, M. and M.B. Jones, 1995. CO_2 and O_2 transport in the aerenchyma of *Cyperus papyrus* L. Aquatic Botany, 52: 93-106.

Maclean, I., R. Boar and C. Lugo, 2011. A Review of the Relative Merits of Conserving, Using, or Draining Papyrus Swamps. Environmental Management, 47: 218-229.

Mannino, I., D. Franco, E. Piccioni, L. Favero, E. Mattiuzzo and G. Zanetto, 2008. A Cost-Effectiveness Analysis of Seminatural Wetlands and Activated Sludge Wastewater-Treatment Systems. Environmental Management, 41: 118-129.

Mburu, N., S. Tebitendwa, D. Rousseau, J. van Bruggen and P. Lens, 2013. Performance Evaluation of Horizontal Subsurface Flow–Constructed Wetlands for the Treatment of Domestic Wastewater in the Tropics. Journal of Environmental Engineering, 139: 358-367.

Meyerson, L.A., K. Saltonstall, L. Windham, E. Kiviat and S. Findlay, 2000. A comparison of *Phragmites australis* in freshwater and brackish marsh environments in North America. Wetlands Ecology and Management, 8: 89-103.

Michael, J., 1983. Papyrus: A new fuel for the third world. Magazine, 99: 418-421.

Mnaya, B., T. Asaeda, Y. Kiwango and E. Ayubu, 2007. Primary production in papyrus (*Cyperus papyrus* L.) of Rubondo Island, Lake Victoria, Tanzania. Wetlands Ecology and Management, 15: 269-275.

Molle, A. Liénard, C. Boutin, G. Merlin and A. Iwema, 2005 How to treat raw sewage with constructed wetlands: an overview of the French systems. Water Science & Technology 51: 11–21.

Morgan, J.A., J.F. Martin and V. Bouchard, 2008. Identifying plant species with root associated bacteria that promote nitrification and denitriefication in ecological treatment systems. Wetlands, 28: 220-231.

Morrison, E., C. Upton, K. Odhiambo-K'oyooh and D. Harper, 2012. Managing the natural capital of papyrus within riparian zones of Lake Victoria, Kenya. Hydrobiologia, 692: 5-17.

Mugisha, P., F. Kansiime, P. Mucunguzi and E. Kateyo, 2007. Wetland vegetation and nutrient retention in Nakivubo and Kirinya wetlands in the Lake Victoria basin of Uganda. Physics and Chemistry of the Earth, Parts A/B/C, 32: 1359-1365.

Muthuri, F.M. and M.B. Jones, 1997. Nutrient distribution in a papyrus swamp: Lake Naivasha, Kenya. Aquatic Botany, 56: 35-50.

Muthuri, F.M., M.B. Jones and S.K. Imbamba, 1989. Primary productivity of papyrus (Cyperus papyrus) in a tropical swamp; Lake Naivasha, Kenya. Biomass, 18: 1-14.

NAS-NRC, 1976. Making aquatic weeds useful: Some Perspectives for Developing Countries. National Academy of Sciences Washington, D.C. 1976.

Neralla, S., R.W. Weaver, B.J. Lesikar and R.A. Persyn, 2000. Improvement of domestic wastewater quality by subsurface flow constructed wetlands. Bioresource Technology, 75: 19-25.

Nyakang'o, J.B. and J.J.A. van Bruggen, 1999. Combination of a well functioning constructed wetland with a pleasing landscape design in Nairobi, Kenya. Water Science and Technology, 40: 249-256.

Okurut, T.O., 2000. A Pilot Study on Municipal Wastewater Treatment Using a Constructed Wetland in Uganda. PhD dissertation, UNESCO-IHE, Institute for Water Education, Delft, The Netherlands.

Osumba, J.J.L., J.B. Okeyo-Owuor and P.O. Raburu, 2010. Effect of harvesting on temporal papyrus (*Cyperus papyrus*) biomass regeneration potential among swamps in Winam Gulf wetlands of Lake Victoria Basin, Kenya. Wetlands Ecology and Management, 18: 333-341.

Perbangkhem, T. and C. Polprasert, 2010. Biomass production of papyrus (Cyperus papyrus) in constructed wetland treating low-strength domestic wastewater. Bioresource Technology, 101: 833-835.

Piedade, M.T.F., W. J. J. , Long,S. P., 1991. The Productivity of the C_4 Grass *Echinochloa polystachya* on the Amazon Floodplain. Ecology, 72: 1456-1463

Rousseau, D., P.L, P.A. Vanrolleghem and N.D. Pauw, 2004. Constructed wetlands in Flanders: a performance analysis. Ecological Engineering, 23: 151-163.

Sala, O.E., Austin, A.T., 2000. Methods of estimating aboveground net primary productivity. In Methods in Ecosystem Science (eds Sala OE, Jackson RB, Mooney HA, Howarth RW), pp. 31–43. Springer-Verlag, New York.

Saunders, M. Jones and F. Kansiime, 2007. Carbon and water cycles in tropical papyrus wetlands. Wetlands Ecology and Management, 15: 489-498.

Saunders and J. Kalff, 2001. Denitrification rates in the sediments of Lake Memphremagog, Canada-USA. Water Research, 35: 1897-1904.

Sekomo, C.B., 2012. Development of a low-cost alternative for metal removal from textile wastewater. PhD Thesis. UNESCO-IHE Institute for Water Education. Delft, The Netherlands.

Sekomo, C.B., Nkuranga, E., Rousseau, D. P. L., Lens, P.N. L., 2010. Fate of Heavy Metals in an Urban Natural Wetland: The Nyabugogo Swamp (Rwanda). Water Air Soil Pollut, 214:321–333

Serag, M.S., 2003. Ecology and biomass production of Cyperus papyrus L. on the Nile bank at
Damietta, Egypt. Mediterranean Ecology, 4: 15-24.

Shetty, U.S., Sonwane, K. D., Kuchekar, S. R., 2005. Water Hyacinth (*Eichornia crassipes*) as a Natural Tool for Pollution Control. Annali di Chimica, 95: 721-725.

Singer, A., A. Eshel, M. Agami and S. Beer, 1994. The contribution of aerenchymal CO_2 to the photosynthesis of emergent and submerged culms of *Scirpus lacustris* and *Cyperus papyrus*. Aquatic Botany, 49: 107-116.

Sonavane, P.G., Munavalli, G. R., Ranade, S. V., 2008. Nutrient Removal by Root Zone Treatment Systems: A Review. Journal of Environmental Science & Engineering, 50: 241-248.

Stein, O.R., J.A. Biederman, P.B. Hook and C. Allen, 2006. Plant species and temperature effects on the k-C* first-order model for COD removal in batch-loaded SSF wetlands. Ecological Engineering, 26: 100-112.

Stottmeister, U., A. Wießner, P. Kuschk, U. Kappelmeyer, M. Kästner, O. Bederski, R.A. Müller and H. Moormann, 2003. Effects of plants and microorganisms in constructed wetlands for wastewater treatment. Biotechnology Advances, 22: 93-117.

Tanner, C.C., 1996. Plants for constructed wetland treatment systems. A comparison of the growth and nutrient uptake of eight emergent species. Ecological Engineering, 7: 59-83.

Terer, T., L. Triest and A. Muthama Muasya, 2012. Effects of harvesting *Cyperus papyrus* in undisturbed wetland, Lake Naivasha, Kenya. Hydrobiologia, 680: 135-148.

Thompson, K., Shewry, P. R., Woolhouse, H. W., 1979. Papyrus swamp development in the Upemba Basin, Zaïre: studies of population structure in *Cyperus papyrus* stands. Botanical Journal of the Linnean Society, 78: 299-316.

van Dam, A., A. Dardona, P. Kelderman and F. Kansiime, 2007. A simulation model for nitrogen retention in a papyrus wetland near Lake Victoria, Uganda (East Africa). Wetlands Ecology and Management, 15: 469-480.

van Dam, A., J. Kipkemboi, F. Zaal and J. Okeyo-Owuor, 2011. The ecology of livelihoods in East African papyrus wetlands (ECOLIVE). Reviews in Environmental Science and Biotechnology, 10: 291-300.

Vymazal, 1999. Removal of BOD_5 in constructed wetlands with horizontal sub-surface flow: Czech experience. Water Science and Technology, 40: 133-138.

Vymazal, 2007. Removal of nutrients in various types of constructed wetlands. Science of the Total Environment, 380: 48-65.

Vymazal, 2011. Plants used in constructed wetlands with horizontal subsurface flow: a review. Hydrobiologia: 1-24.

Vymazal and L. Kröpfelová, 2009. Removal of organics in constructed wetlands with horizontal sub-surface flow: A review of the field experience. Science of the Total Environment, 407: 3911-3922.

Yang, Q., Z.-H. Chen, J.-G. Zhao and B.-H. Gu, 2007. Contaminant Removal of Domestic Wastewater by Constructed Wetlands: Effects of Plant Species. Journal of Integrative Plant Biology, 49: 437-446.

Zhu, S.-X., H.-L. Ge, Y. Ge, H.-Q. Cao, D. Liu, J. Chang, C.-B. Zhang, B.-J. Gu and S.-X. Chang, 2010. Effects of plant diversity on biomass production and substrate nitrogen in a subsurface vertical flow constructed wetland. Ecological Engineering, 36: 1307-1313.

Zurita, F., M.A. Belmont, J. De Anda and J. Cervantes-Martinez, 2008. Stress detection by laser-induced fluorescence in Zantedeschia aethiopica planted in subsurface-flow treatment wetlands. Ecological Engineering, 33: 110-118.

Chapter 3: Performance evaluation of horizontal subsurface flow constructed wetlands for the treatment of domestic wastewater in the tropics

This chapter has been published as: Mburu, N., Tebitendwa, S., Rousseau, D., van Bruggen, J., Lens, P., 2013. Performance Evaluation of Horizontal Subsurface Flow–Constructed Wetlands for the Treatment of Domestic Wastewater in the Tropics. Journal of Environmental Engineering 139(3) 358-367.

Abstract

The lack of information on constructed wetland performance in the tropics is among the factors that have hindered the adoption of low-cost natural wastewater treatment technologies as alternative to conventional wastewater treatment. A pilot scale study was undertaken in Juja (Kenya) to assess the performance of horizontal subsurface flow constructed wetlands (HSSF-CWs) under tropical conditions. Primary domestic wastewater effluent was continuously fed into three replicate wetland cells each with an area of 22.5 m^2 (7.5 x 3 m) and with gravel as substrate. The study revealed successful performance of the wetlands in terms of compliance with local discharge standards with respect to COD, BOD$_5$, TSS and SO$_4$-S at a percentage average mass removal efficiency between 58.9% and 74.9%. Moderate removal of NH$_4^+$-N and TP were recorded. The estimated first order aerial-rate constant and the BOD$_5$ background concentration showed the HSSF-CW to be area requirement competitive. The good performance in organic matter and suspended solids removal reveals that HSSF-CW can help to alleviate the current environmental pollution problems experienced in developing countries due to the discharge of partially treated or untreated domestic wastewater.

3.0   Introduction

One of the commonly encountered environmental problems in developing countries is water pollution caused by direct disposal of untreated or partially treated wastewater (Mara, 2004; Tsagarakis et al., 2001). The discharge of polluted wastewater from households and industries is a threat to nature and humans in these low and middle income economies because it causes eutrophication of surface waters and transmission of water-borne diseases. Often the reason for the lack of wastewater treatment is financial, but it is also due to the lack of the application of low-cost wastewater treatment technologies.

For developing countries, short-term and medium-term solutions lie in the use of cheap and robust wastewater treatment technologies (Okurut, 2000). Constructed wetlands, particularly horizontal subsurface flow systems, are well established for the priority treatment of domestic wastewater (Wiessner et al., 2005). Constructed wetland utilization can provide sustainable wastewater treatment: because they rely on natural processes, they are less expensive to build, operate and maintain compared to conventional sewage treatment systems. The purified water is suitable for re-use and harvested plants can have economic value. All these are benefits for the application this wastewater treatment technology in developing countries facing social-economic challenges.

The use of this technology is unfortunately still limited in developing countries due to poor understanding of the constructed wetland potential (Diemont, 2006; Kivaisi, 2001; Mashauri et al., 2000). This is caused by a lack of information on constructed wetland performance in tropical regions (Bojcevska and Tonderski, 2007). Nonetheless, it can be assumed that constructed wetlands are even more suitable for wastewater treatment in tropical than in temperate areas because a warm climate is conducive to year-round plant growth and microbiological activity, which in general have a positive effect on a wetland's treatment efficiency (Bojcevska and Tonderski, 2007; Haberl, 1999; Kivaisi, 2001).

In the developed, temperate climate countries, constructed wetlands with horizontal sub-surface flow (HSSF-CW) have been successfully used for treatment of various types of wastewater for more than four decades (Vymazal and Kröpfelová, 2009). Their application has been driven by a rising cost of energy associated with conventional wastewater treatment systems and increasing concerns about climate change which provide a financial incentive, as well as public support, to the implementation of this low energy consumption 'green' technology. (Lee et al., 2009). In this sense the constructed wetland wastewater technology is deemed a potential alternative to reduce both greenhouse gas emissions and power consumption.

Most systems have been designed to treat municipal sewage and are generally efficient in removal of organic matter (BOD) and suspended solids (SS), but the removal of nitrogen and phosphorus is often relatively poor (Tanner et al., 1999; Verhoeven and Meuleman, 1999; Vymazal, 1996, 2005). The degradation of wastewater contaminants within these systems takes place through a number of physical, chemical and biological processes that occur simultaneously, and is enhanced by the interactions between water, granular media, macrophytes, litter, detritus and microorganisms. Organic matter reduction in HSSF-CW is mainly achieved by microorganisms attached to the substrate media and to plant roots through aerobic, anoxic and anaerobic processes (Faulwetter et al., 2009), while its removal efficiency is influenced by the loading rate, residence time, which is a function of bed volume and flow rate; plant type and environmental factors including temperature (Kadlec and Wallace, 2009). Nitrogen undergoes several transformations in wetlands, including ammonification, nitrification, denitrification, adsorption, bacteria and plant uptake (Vymazal, 2007). Plants in wetlands play a major role in providing additional media onto which microorganisms can attach, help maintain aerobic microsites in the wetland bed through oxygen transfer via roots and rhizome systems, and control growth of algae by restricting sunlight penetration (Tanner, 2001). A recent state-of-the-art review and discussion on contaminant removal processes in subsurface-flow constructed wetlands is provided by García et al. (2010).

In Kenya, sustainable wastewater management has not yet been achieved as a result of population growth, rapid urbanization and a surge towards a higher standard of living in the context of economic constraints (AMCOW, 2006; Nzengy'a and Wishitemi, 2001). Although substantial progress has been made in the provision of services for collection, treatment and disposal of wastewater in urban areas, sanitation infrastructure in non-urban areas is characterized by low levels of access, inadequate or non-functioning wastewater treatment facilities (WB-WSP, 2012). The estimated sewer coverage and connection rate is between 12-19 % and only 5% of the national sewerage is effectively treated due to low operational capacity of utilities (Pokorski and Onyango, 2010). Discharge of untreated or partially treated domestic wastewater in the city suburbs, mushrooming townships, remotely located institutions and industries, and rural areas, have been identified as major causes of fresh water resource contamination and degradation in the form of deteriorating water quality of groundwater, rivers and lakes, and the spread of waterborne diseases (Mogaka et al., 2006; Pokorski and Onyango, 2010). Difficulties involved in securing finances to solve this problem has brought the need to search for cheaper and appropriate solutions suitable to Kenyan conditions (Abira, 2007; Bojcevska and Tonderski, 2007; Nyakang'o and van Bruggen, 1999).

Like other developing tropical countries, the need for appropriate low-cost wastewater treatment technology in Kenya is evident. In particular a noticeable difference could be achieved by ensuring improvement of domestic sewage effluent before discharge into natural water courses. Nyakang'o and van Bruggen (1999), Nzengy'a et al. (2001), Bojcevska and Tonderski (2007) and Abira (2007), showed that the use of constructed wetlands can help alleviate the problems of discharging partially or untreated wastewater into aquatic systems. For Kenya to meet the current wastewater treatment requirements and achieve the millennium development goal of clean safe water and sanitation by 2015, the option of constructed wetlands should be considered as a sustainable technology for improving effluent quality. Thus, the objective of the present study was to assess the removal capacity of a pilot scale HSSF-CW planted with the tropical macrophyte *Cyperus papyrus* as a low-cost technology for treating domestic sewage. We used an existing pilot scale HSSF-CW built in 2003, at the Jomo-Kenyatta University of Agriculture and Technology (JKUAT), Juja (Kenya) receiving a continuous gravity feed of primary effluent from a facultative pond treating domestic wastewater (Kibetu, 2008). This study seeks to contribute performance data and information for the HSSF-CW with *Cyperus papyrus* macrophyte performing secondary treatment of domestic wastewater in the tropics. To test the influence of the macrophyte native *Cyperus papyrus,* an unplanted control unit was included in the set-up.

3.1 Methodology

3.1.1 Study area

The pilot scale HSSF-CW was sited within the JKUAT sewage treatment works, 40 km north east of Nairobi city (Kibetu, 2008). The area is at an altitude of 1463 m above sea level at coordinates 1°05'45" S and 37°1'25" E. JKUAT sewage works treats domestic wastewater by use of a set of 5 wastewater stabilization lagoons: 2 primary facultative ponds and 2 secondary facultative ponds in parallel and 1 maturation pond in series to all. The mean monthly temperature, precipitation and potential evaporation (Eo) values for Juja area are given in Table 3.1.

Table 3.1. Mean monthly temperature and precipitation for Juja area (Data from Wanjogu and Kamoni, 1986)

Month	Jan	Feb	Mar	Apr	May	Jun	Jul	Aug	Sep	Oct	Nov	Dec	Annual
Mean Temp. ($^{\circ}$C)	20.3	21	21.3	20.7	19.9	19	18.4	18.6	19.4	20.1	20.1	19.9	19.7
Precipitation (mm)	33	34	99	202	126	34	19	22	21	61	128	77	856
Eo (mm)	124	124	124	99	87	87	74	74	74	124	99	111	1856

3.1.2   Design of the pilot scale HSSF-CW

The pilot scale wetland system consisted of three cells set in parallel, each 22.5 m^2. The macrophyte *Cyperus papyrus* was growing in two of the cells, referred to as Cyp1 and Cyp2, while one cell remained unplanted and acted as the control (Ctrl). *Cyperus papyrus,* a common tropical wetland plant (Okurut, 2000), is locally occurring in the study area. The sizing of the wetland was in accordance with Kadlec and Knight (1996) and the design considerations were based on the total area necessary to remove BOD$_5$. The cells were built in such a way that controlled and measurable quantities of wastewater and rain were the only inputs into the system. The wetland received a continuous gravity feed of primary effluent from the primary facultative pond at the JKUAT sewage works. The desired flow rate of the influent wastewater was maintained manually by regulating a gate valve at the inlet works of the HSSF-CW system, while the water depth was maintained at 0.5 m within the gravel bed with the aid of fixed outlet pipes (collecting effluent from the floor level of the wetland cell).

The wetland cells were 7.5 m long and 3 m wide with vertical masonry sides, 0.95 m deep, and a concrete floor sloped at one percent. The cells were filled with granite type gravel to a depth of   0.6 m, ranging in size from 9-37 mm, with a porosity of 45 %. The larger size gravel was placed near the inlet and outlet of the wetland to help with uniform distribution of the influent wastewater stream and drainage of the wetland, respectively. The macrophyte *Cyperus papyrus* was established into two of these cells (Cyp1 and Cyp2), using rhizome fragments at a spacing of 0.75 m by 0.75 m. Routine maintenance involved weeding the HSSF-CW bed, grass cutting around the pilot HSSF-CW site and cleaning of the inlet works of large floating solids. Non-routine maintenance work involved patching the masonry with plaster mortar to stop detected leaks in the wetland masonry.

3.1.3 Monitoring of wetland performance and standing biomass

Data sets for the performance of the pilot HSSF-CW were obtained over the period October 2008 - January 2011. All wastewater samples were grab samples taken manually. Sampling and sample handling was as given in the Standard Methods and Procedures for the Examination of Water and Wastewater (APHA, 1998). Samples were taken at the influent and effluent points of the pilot scale HSSF-CW cells, while simultaneously determining the flow rate by the volumetric method. The parameters measured to assess the performance of the HSSF-CW included COD, BOD$_5$, TSS, NO$_3^-$-N, NH$_4^+$-N, TP, PO$_4^{3-}$-P, SO$_4^{2-}$-S.

Environmental and some physical-chemical parameters were measured in-situ at the inlet and outlet of the HSSF-CW, respectively, for the influent wastewater and the treated effluent. pH, electrical conductivity (EC) and total dissolved solids (TDS) were determined with a HACH ECO 40 multi-probe, whereas dissolved oxygen and temperature were measured using a dissolved oxygen meter; model HACH HQ40d multi probe.

Standing biomass was estimated monthly by counting the number of shoots within a 1 m^2 quadrant at the mid-section of the wetland bed and cross-checking between the two planted cells. The numbers obtained were used to compute the plant shoot density per unit area.

3.1.4 Data analysis

Characterization of the HSSF-CW performance was achieved by computation of input (m_i) and output (m_o) mass loading rates (as the product of concentration and influent/ effluent flow rate respectively) for each individual constituent at sampling events. Mass removal rates were calculated as a difference between input and output mass loading rates. Percentage mass removal for each constituent was calculated as mass removal $(\%) = 100 \times \dfrac{m_i - m_o}{m_i}$. The hydraulic loading rates were calculated based on the wastewater inflow rates while the hydraulic retention times were based on the averages between the inflow and outflow rates. Statistical analysis was performed using MINITAB 15 and Microsoft Excel 2007 software. Comparison of variables was performed using the analysis of variance technique (ANOVA), the Tukey's multiple comparisons was used to test differences in the means and the Anderson-Darling test was used for testing normality. First-order area-based removal rate constants (k), assuming exponential removal to non-zero background concentrations (C*), were estimated for COD and BOD$_5$ removal. Fitted values of k and C* were derived from the following equation (Kadlec and Knight, 1996): $\ln\dfrac{C_O - C^*}{C_i - C^*} = \dfrac{-k}{q}$ where k is the area-based first-order removal rate constant (m/d), q is the hydraulic loading rate (m /d), C_o is the outlet concentration (mg/ l), C_i is the inlet concentration (mg/ l), and C^* is the irreducible background concentration in the wetland. The fittings were performed using the non-linear regression procedure of Statgraphics Centurion XVI version 16.1.11.

3.2 Results

3.2.1 General treatment performance

The results obtained from the laboratory analysis of influent-effluent wastewater samples, together with the environmental parameters, and the wastewater discharge guidelines to surface water courses in Kenya, according to NEMA (2003), are summarized in Table 3.2. The concentrations of analyzed parameters in the influent varied considerably during the study period. They were significantly higher than in the effluent except for TP, PO_4^{3-}-P, NO_3^--N, and NH_4^+-N. The influent wastewater to the pilot scale HSSF-CW can be classified as weak strength domestic wastewater in terms of BOD$_5$, COD, TSS, TDS, TP and PO_4^{3-}-P; medium strength in terms of NH_4^+-N and strong in terms of SO_4^{2-}-S according to Metcalf and Eddy (2003).

Analysis of the performance of the pilot scale HSSF-CW system is presented in Table 3.3. The warm temperatures experienced in the study periods are typical for tropical climates, with the measured water temperatures of the influent ranging between 19.4 °C and 26.6°C. The range of influent wastewater pH was 7.22 - 8.67, while that of the effluent wastewater from the HSSF-CW system remained rather circum-neutral. Low values of below 2.0 mg/L dissolved oxygen were measured from the pilot HSSF-CW effluent (Cyp1: 0.65-0.95 mg/L, Cyp2: 0.31-1.06 mg/L, Ctrl: 0.53-1 mg/L), while the influent wastewater registered between 0.2-18.5 mg/L dissolved oxygen. An apparent increase in TDS concentration and EC in the effluent of the

HSSF-CW system was determined to be not statistically significant: TDS (Ctrl, p=0.155; Cyp1, p=0.174; Cyp2, p=0.092) and EC (Ctrl, p=0.163; Cyp1, p=0.181; Cyp2, p=0.094).

3.2.2   Plant growth and HSSF-CW maintenance

The *Cyperus papyrus* macrophyte established well without the need for soil on the gravel-based HSSF-CW cells loaded with primary effluent of domestic wastewater (Fig. 3.1a). A progressive increase in plant density, shoot length and stem diameter was observed. The macrophyte grew vigorously from about 2-3 rhizhome fragments m^{-2} to an average of 100 shoots m^{-2} in eight months (Fig.3.2), with a height up to 3 m, and formed a dense stand covering the wetland surface extensively (Fig.3.1b). Senescence and re-growth of above ground plant parts occurred concurrently during the study period. During a renovation exercise to fix water leakage, the roots of the macrophytes were found to have penetrated within the gravel bed up to a depth of 0.4 m (Fig.3.1c). The macrophyte regenerated successfully from rhizomes on three occasions after harvesting but performed poorly on the fourth occasion and had to be re-established with rhizome propagules.

Table 3.2. Influent–Effluent Wastewater Characteristics in the Study Period

| Parameter | Unit | Influent | | | Effluent (Mean ± st.dev.) | | | | | | Discharge standards |
		Range	Mean ± st.dev.	n	Ctrl	n	Cyp1	n	Cyp2	n	
DO	mgL⁻¹	0.21-18.51	6.32 ± 0.6	9	0.75 ± 0.14	9	0.85 ± 0.12	5	0.72 ± 0.22	5	
Temp	°C	19.4-26.6	23.3 ± 1.22	10	22.8 ± 1.7	9	23.2 ± 1.7	5	22.6 ± 0.8	5	40
pH	pH	7.22-8.67	7.8 ± 0.2	10	7.2 ± 0.1	9	7.04 ± 0.0	5	7.11 ± 0.11	5	5 - 9
EC	µScm⁻¹	242-319	280.1 ± 7.9	10	296.0 ± 27.3	9	290.5 ± 15.9	5	301.3 ± 31.3	5	
TDS	mgL⁻¹	155-204	179.1 ± 5.1	10	189.6± 17.7	9	189.7± 10.0	5	192.8 ± 19.6	5	
TP	mgL⁻¹	2.3-4.8	3.6 ± 0.2	9	3.3 ± 0.2	8	2.2 ± 0.6	5	2.8 ± 0.3	5	
PO_4^{3-}-P	mgL⁻¹	1-2.4	1.4 ± 0.1	10	1.4 ± 0.2	10	0.6 ± 0.2	7	0.8 ± 0.3	7	1
NO_3^--N	mgL⁻¹	0.1-3	1.1 ± 1.1	21	0.8 ± 0.8	21	1.1 ± 0.8	16	0.9 ± 0.9	19	10
NH_4^+-N	mgL⁻¹	18-33	25.8 ± 4.5	19	20.5 ± 5.3	17	18.8 ± 3.2	13	19.0 ± 5.8	19	1
TSS	mgL⁻¹	20-430	103.1 ± 104.1	26	40.1 ± 25.1	25	27.9 ± 16.6	20	25.5 ± 17.5	24	35
BOD₅	mgL⁻¹	12-105	73.6 ± 17.7	10	38.9 ± 13.3	10	34.6 ± 12.3	7	28.9 ± 9	9	40
COD	mgL⁻¹	52-354	159.5 ± 75.8	48	105.0 ± 64.3	22	89.5 ± 45.1	22	91 ± 31.8	46	120
SO_4^{2-}	mgL⁻¹	18.57-100	66.7 ± 25.4	20	22.0 ± 14.3	20	29.3 ± 12.2	17	20.1 ± 16.2	19	750

Table 3.3. The mean overall performance of the HSSF-CW system evaluated in terms of applied surface loading rates (LR), observed removal rates (RR) and percentage mass removal efficiency

| Parameter | Ctrl (Mean ± st.dev.) | | | Cyp1 (Mean ± st.dev.) | | | Cyp2 (Mean ± st.dev.) | | |
	LR (gm⁻²d⁻¹)	RR (gm⁻²d⁻¹)	% Removal	LR (gm⁻²d⁻¹)	RR (gm⁻²d⁻¹)	% Removal	LR (gm⁻²d⁻¹)	RR (gm⁻²d⁻¹)	% Removal
COD	24.9 ± 13.6	14.9 ± 9.7	60.1 ± 22.4	28.4 ± 9.8	17.9 ± 6.4	65.0 ± 17.2	26.1 ± 14.6	14.9 ± 10.4	59.2 ± 18.8
TSS	15.4 ± 11.7	9.2 ± 9.9	44.2 ± 30.1#	17.8 ± 15.3	12.2 ± 11.4	58.9 ± 15.6	16.9 ± 14.4	13.0 ± 13	62.0 ± 25.5
BOD₅	12.7± 7.0	3.6 ± 1.9	59.7 ± 19.0	12.9 ± 10.0	9.6 ± 8.2	69.6 ± 8.4	10.4 ± 8.9	5.5 ± 1.2	67.3 ± 8.8
NH_4^+-N	4.6 ± 2.7	1.8 ± 0.8	44.1 ± 15.6	4.4 ± 1.7	2.1 ± 0.43	46.6 ± 1.0	5.6 ± 3.5	2.7 ± 1.9	49.8 ± 18.0
SO_4^{-2}-S	10.5 ± 7.8	8.1 ± 7.3	71.9 ± 26.0	14.7 ± 12.4	11.3 ± 10.4	66.2 ± 26.5	11.2 ± 9.9	8.9 ± 8.6	74.9 ± 25.4
TP	0.71 ± 0.26	0.22 ± 0.1	32.9 ± 4.4#	0.7 ± 0.3	0.5 ± 0.2	49.6 ± 15.5	0.5 ± 0.2	0.24 ± 0.1	47.9 ± 6.15
PO_4^{3-}-P	0.18 ± 0.1	-0.03 ± 0.02	-19.1 ± 15.6#	0.2 ± 0.1	0.1 ± 0.1	57.1 ± 19.2	0.13 ± 0.03	0.05 ± 0.02	42.0 ± 19.2

#: Statistically significant difference with Cyp1 and Cyp2 (α=0.05)

Fig. 3.1: Planted pilot wetland cells showing (a) successful establishment of tropical macrophyte *Cyperus papyrus* in the pilot gravel beds, (b) progressive increase in plant density, shoot length and stem diameter and (c) root penetration within the gravel bed

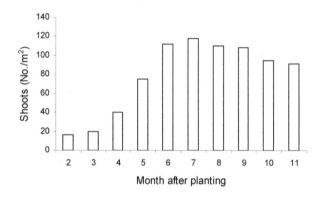

Fig.3.2. Variation of *Cyperus papyrus* shoot density over the period October 2008 (start-up) to July 2009 (harvest)

3.2.3 Water balance

The mean hydraulic loading rates during the study period were 195 ± 80 mm/d, 193 ± 86 mm/d and, 176 ± 90 mm/d for the wetland cells Cyp1, Cyp2 and Ctrl, respectively, yielding mean theoretical hydraulic retention times of 1.6 ± 1.4 days in Cyp1, 1.95 ± 1.0 days in Cyp2 and 1.96 ± 0.8 days in the Ctrl cell. The difference in hydraulic loading rates in the course of the study was found not to be statistically significant (Cyp1& Ctrl, p=0.117; Cyp1 & Cyp2, p=0.103; Cyp2 & Ctrl, p=0.907). The mean water loss (and range) from the pilot wetland cells amounted to 19.8 ± 10.4 mm/d (4.0-30.4 mm/d), 13.6 ± 8.4 mm/d (4.4-31.4 mm/d) and 8.6 ± 8.4 mm/d (0-17.8 mm/d) in Cyp1, Cyp2 and Ctrl, respectively, based on direct inflow-outflow measurements. Evapotranspiration rates from the individual planted cells were not significantly different from each other (p=0.559). However, the water loss in the control cell was significantly different from the planted cell (Ctrl & Cyp1: p=0.033, Ctrl & Cyp2: p=0.012).

3.2.4 Temperature and pH

The influent wastewater temperature range (19.4-26.6 $^{\circ}$C) correlated well with the ambient air temperature averages (Table 3.1). Effluent wastewater temperatures for the pilot HSSF-CW cells were significantly lower (0.5-3.5°C) in the effluent than influent (p<0.05), with an effluent wastewater temperature range of 17.2-25.9 $^{\circ}$C in Cyp1, 17.2-24.2 $^{\circ}$C in Cyp2 and 17.3-26.2 $^{\circ}$C in Ctrl. Side by side comparison showed no significant differences in effluent temperature among the pilot cells (Cyp1& Ctrl, p=0.696; Cyp1 & Cyp2, p=0.412; Cyp2 & Ctrl, p=0.774).

The outflow wastewater pH was circum-neutral (Cyp1: 7.01-7.47; Cyp2: 7.12-7.42; Ctrl: 6.99-7.32) and dropped 0.55-0.69 pH units in the pilot HSSF-CW cells. The influent and effluent pH was determined significantly different only for the planted cells (Influent & Cyp1, p=0.028; Influent & Cyp2=0.03; Influent & Ctrl, p=0.09)

3.2.5 Performance

3.2.5.1 COD and BOD$_5$

The BOD$_5$:COD ratio in the influent wastewater to the pilot HSSF-CW ranged between 0.13 - 0.83, indicating that the primary effluent from the facultative pond had a variable biodegradability over the course of the study. Table 3.3 gives the COD and BOD$_5$ surface loading rates (LR), removal rates (RR) and removal efficiencies for the three cells. Statistical analysis showed there was no significant difference in the organic matter loading rates among the cells (Cyp1 & Ctrl, p = 0.194; Cyp1& Cyp2, p = 0.169; Cyp2 & Ctrl, p = 0.930). The organic matter removal efficiency varied from time to time in the same cell as a result of the variation in surface loading rate and residence time (Fig.3.3).

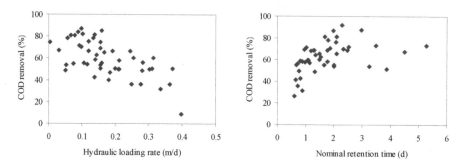

Fig. 3.3. Effect of hydraulic loading rate and hydraulic retention time on COD removal in the planted cells, Cyp1 and Cyp2 (pooled data)

COD and BOD_5 mass removal efficiencies ranged between 24-74.7% and 61.5 - 80.9% in Cyp1, 31.5-87.7% and 44.2 - 91.8% in Cyp2, 15-81% and 36.9 - 90.6% in the Ctrl cell, respectively. No significant difference was observed in the removal efficiencies for both COD and BOD_5 between the pilot HSSF-CW cells (p>0.05).

The relationship between the COD loading and removal rates for the three cells is shown in Fig.3.4. Figure 3.4 shows the relationship was strong (R^2= 0.70-0.92) and linear, with removal rates increasing as the loading rates increased. However, beyond a COD loading rate of approximately 20 g $m^{-2}d^{-1}$ and 30 g $m^{-2}d^{-1}$ in the Ctrl and planted cells, respectively, the relationship appears rather variable (data points do not lie close to the regression line).

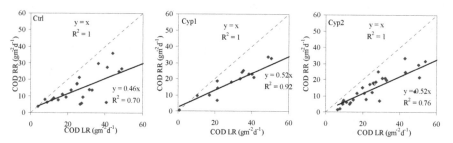

Fig. 3.4. Linear regression analysis of observed mass removal rates on mass loading rates for COD

During the whole study period, effluent concentrations with zero mg/l BOD_5 or COD were not obtained from the pilot HSSF-CW cells. The residual background concentration or the amount of organic matter produced by the wetland system itself was estimated for the planted cells using the first order k-C* equation as described in Kadlec and Knight (2006). The estimated first order aerial-rate constant (k) and the background concentration (C*) for COD and BOD$_5$, were Cyp1 (k=0.31 m/d, C*=70.3 mg/L), Cyp2 (k=0.35, C*=82.3 mg/L) and Cyp1 (k=0.10 m/d, C*=10.0 mg/L), Cyp2 (k=0.169, C*=17.1 mg/L), respectively.

3.2.5.2 TSS

Only the planted HSSF-CW cell Cyp2, showed a better TSS removal in terms of effluent quality (Table 3.2), compared to the unplanted Ctrl cell (p=0.007). The difference in TSS effluent concentrations was not significant between the Cyp1 and Ctrl cells (p= 0.087). TSS mass removal efficiencies ranging between 32.3-86.1%, 25-96.8 %, and 0-95.5% were achieved in Cyp1, Cyp2 and Ctrl cells of the HSSF-CW pilot cells, respectively. The difference in the TSS removal rates (Table 3.3) among the cells was not significant (Cyp1 &Ctrl, p= 0.651; Cyp1 &Cyp2, p=0.668; Cyp2 &Ctrl, p=0.300). The mass removal rates showed a strong positive linear relationship with the mass loading rates in three cells (Fig. 3.5). Figure 3.5 shows that the relationship was variable beyond a TSS loading rate of about 10 g m^{-2}d^{-1} and 20 g m^{-2}d^{-1}, in the Ctrl and planted cells, respectively.

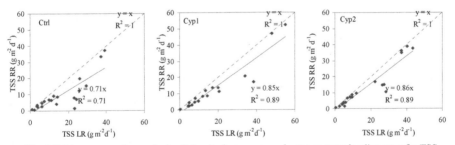

Fig. 3.5. Linear regression analysis of observed mass removal rates on mass loading rates for TSS

3.2.5.3 Nitrogen

Discrepancies between inlet and outlet measurements of NH$_4^+$-N and NO$_3^-$-N were evaluated. Moderate mean mass NH$_4^+$-N removal efficiencies ranging between 21.2-53.6 %, 20.1-72.5 % and 14.4-60.1 % were observed in Cyp1, Cyp2 and Ctrl cell of the HSSF-CW, respectively. There was no significant difference in the NH$_4^+$-N polishing rates of 2.1±0.43 g m^{-2} d^{-1}, 2.7±1.9 g m^{-2} d^{-1} and 1.8±0.8 g m^{-2} d^{-1}

determined in the Cyp1, Cyp2 and Ctrl cells of the HSSF-CW, respectively (Cyp1 &Ctrl, p=0.730; Cyp1&Cyp2, p=0.668; Cyp2&Ctrl, p=0.239). In the course of the study, the nitrate concentration both in the influent and effluent did not exceed 3 mg/L and in some cases was non-detected (Cyp1: 0.12-2.31 mg/L, Cyp2: 0.11-2.58 mg/L, Ctrl: 0-2.8 mg/L); at the same time not showing a significant difference between the influent and effluent concentration (Cyp1, p=0.813; Cyp2, p=0.545 and Ctrl, p=0.767).

3.2.5.4 Phosphorus

Modest TP mass removal efficiencies ranging between 29.2-42.7 %, 30-63.3 5% and 39.7-57.4% were achieved in the Ctrl, Cyp1 and Cyp2 cells, respectively. The effluent TP concentration in the planted cells Cyp1 (1.3-2.8 mg/L) and Cyp2 (2.3-3.1 mg/L) did not have a significant difference (p=0.073), whereas the effluent TP concentration from the unplanted cell, Ctrl (2.9-3.5 mg/L) showed a significant difference with the planted cells Cyp1 (p=0.002) and Cyp2 (p=0.000).

Removal of the soluble reactive phosphorus (PO_4^{3-}-P) was only observed in the planted cell at mass efficiencies ranging between 25-76.9% (Cyp1) and 17.1-69.7 % (Cyp2), whereas in the unplanted cell, there was mainly an increase of PO_4^{3-}-P effluent concentration with removal efficiencies of -35.5 - 3.7%.

3.2.5.5 Sulphates

There was a significant difference in the influent and effluent SO_4^{-2}-S concentration (3.3) in the three cells (p=0.000). A high correlation coefficient ($R^2 = 0.92$-0.95) was determined for the linear relationship between mass SO_4^{-2}-S loading and removal rates in the pilot scale cells (Fig.3.6). The mean removal efficiencies (Table 3.3) of 66.2±26.5% (13.8-93.6%) in Cyp1, 74.9±25.4 (7.7-96.2%) in Cyp2 and 71.9±26 (3.1-96.2%) in Ctrl did not show any significant difference (Cyp1 & Ctrl, p=0.526; Cyp2 & Ctrl, p=0.728; Cyp1 & Cyp2, p=0.329).

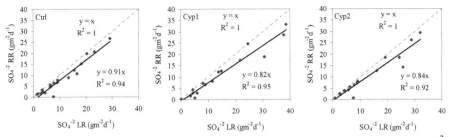

Fig. 3.6. Linear regression analysis of observed mass removal rates on mass loading rates for SO_4^{-2}-S

3.3 Discussion

3.3.1 Environmental factors and estimated water balance

This study contributes performance information for the HSSF-CW performing secondary treatment of domestic wastewater in the tropics. The influent wastewater to the wetland (Table 3.2) is within the influent range classification for secondary treatment of 30-100 mg/L BOD (Kadlec and Wallace, 2009). The fluctuation of the abiotic factors, namely temperature, dissolved oxygen and temperature in the influent wastewater varied according to the outdoor conditions and treatment factors in the upstream facultative lagoon. The mean temperatures recorded for the influent wastewater was above 20°C (Table 3.2) and the temperature range (19.4-26.6 °C) was optimal for removal of nutrients and organic matter (Kadlec and Reddy, 2001).

The observed wide range of dissolved oxygen concentration in the influent to the constructed wetland system (0.2-18.5 mg/L) was attributed to algal growth and photosynthetic activity in the non-shaded environment of the facultative pond. Indeed, the mean influent DO of 6.3±0.6 mg/L is high compared to treatment wetlands that receive wastewater from septic tanks or anaerobic systems. However, the mean dissolved residual oxygen concentrations did not exceed 1.0 mg/L in the effluents of the pilot HSSF-CW cells, suggesting high oxygen consumption among competing aerobic processes within the HSSF-CW bed. Oxygen consumption in this bed is mainly related to the differences between inlet and outlet BOD and ammonia. Subsurface systems that are more heavily loaded with biochemical and nitrogenous oxygen demands have essentially no DO in their effluents (Kadlec and Wallace, 2009).

The difference between the inlet and outlet pH was only significant in the planted units (Table 3.2), suggesting an influence of pH buffering by the macrophytes. The pH results for the constructed wetland cells are consistent with the behavior of wastewater pH in other treatment wetlands, which is circum-neutral unless influents are strongly basic or acidic (Kadlec and Knight, 1996).

The water balance showed that in all the cells, differences in flow existed between inflow and outflow, as a result of physical evaporation and plant transpiration. The evapotranspiration rates for this study are, however, higher than those reported in a fringing papyrus swamp on Lake Naivasha of 12.5 mm/day (Jones and Muthuri, 1985), but within the range of those reported for a subsurface horizontal flow wetland of 24.5±0.6 mm/d in Uganda (Kyambadde et al., 2005). In the course of the study, repairs were undertaken to fix leaks detected in the masonry walls of the HSSF-CW which could have affected the overall water balance estimation.

3.3.2   Removal of COD and BOD$_5$

The BOD$_5$ and COD removals efficiencies are lower than the average values of 85% and 75% removal for different countries reported by Vymazal (2005) but similar to the results obtained by Mashauri (2000) who found 57-74% (at a low infiltration rate) and 42-59% (at a high infiltration rate) reduction in COD by using a 0.75 m depth HSSF-CW planted with *Typha latifolia* reeds. On the other hand, the HSSF-CW showed potential to achieve > 80% COD removal at low hydraulic loading rates (Fig. 3.3). This result coincide with that found by Okurut (2000) who assessed the effect of hydraulic loading rate on organic matter removal in a HSSF-CWs planted with a *Cyperus papyrus*.

Compared to the oxygen demand of the wastewater (12-105 mg BOD$_5$/l), the oxygen concentration in the influent, with a mean of 6.3 ± 0.6 mg/L, is insufficient, especially for the higher range of organic loading. Therefore, a large proportion of the COD and BOD$_5$ removal probably occurred by anoxic and anaerobic process, among others denitrification and sulphate reduction. This is supported by the observation that nitrates did not accumulate (Table 3.2), notwithstanding the modest nitrification, and significant sulphate removal in the pilot HSSF-CW cells (Table 3.3). In this study, the effluent SO_4^{2-}-S concentrations were significantly lower than those in the influent for all the three pilot HSSF-CW cells (Table 3.2), suggesting that sulphate reduction was important. Indeed, sulphate reduction is the most effective biochemical reaction in removing dissolved organic matter measured as COD (36–100% of the total removal) in wetlands with a depth of 0.5 m (Garcia et al., 2004). The observed sulphate removal efficiency in both the planted and unplanted cells, ranging between 66.2 ± 26.5 % to 74.9 ± 25.4 %, agrees well with those reported in literature. For instance, a laboratory scale HSSF-CW test with a median influent SO_4^{2-}-S concentration of 75±16 mg/L, achieved a median sulphate removal efficiency of 71±32% and 79±25% for planted and unplanted set-ups, respectively (Baptista et al., 2003).

The organic matter loading rate had an inverse relationship with the removal efficiency (not shown), highlighting the importance of the organic loading and retention time (Fig.3. 3) on the treatment of organic matter. The organic matter loading rate is a variable that combines the effect of the COD or BOD_5 concentration, wetland area and retention time. The retention time determines the contact between the pollutants and the removal processes. The mean retention time of 1.6 -2.0 days in the pilot wetland cells is actually a little low for a HSSF-CW, and may have contributed to the fair performance of the pilot wetland cells. A plot of organic matter removal versus nominal retention time in Fig. 3.3 suggests a retention time of at least 2 days would be optimum for the system, under the applied experimental conditions.

Although, there was no significant difference in the organic matter removal efficiencies between the unplanted and planted cells, the planted cells were able to sustain a high organic loading rate (up to 30 g m^{-2} d^{-2}), compared to the control cell, while maintaining a linear relationship with the removal rate (Fig. 3.4). This suggests a positive influence of the plants on the wetland performance. It has been proposed that the planted rhizosphere stimulates the microbial community density and activity by providing root surface for microbial growth, carbon sources through root exudates and a micro-aerobic environment via root oxygen release (Baptista et al., 2003; Gagnon et al., 2007; Kadlec and Wallace, 2009).

The outlet concentrations of the planted cells were fitted to the first order k-C* model. The obtained C* values for COD and BOD_5 were within the typical range (10-100 mg/L for COD and 1-10 mg/L for BOD_5) reported by Kadlec and Wallace (2009), except for cell Cyp2 where the C* value for BOD_5 was 17.1 mg/L. It was, however, within the range cited by Okurut (2000) for pilot wetlands in Uganda, planted with *Cyperus papyrus* (12 $mgBOD_5$/L) and *Phragmites mauritianus* (17$mgBOD_5$/L). The non-zero background levels of both BOD_5 and COD processed by wetlands could be of importance to environmental regulators, as it tends to limit the extent of organic matter removal that can be obtained in treatment wetlands. The land area requirement for secondary treatment (based on BOD_5 removal) was estimated as 2.0 m^2 per population equivalent (p.e), estimated from the obtained area-based first order rate constant for BOD_5 removal k = 0.1 m d^{-1}, a BOD contribution assumed at 50 g p.e^{-1} d^{-1}, a wastewater flow rate of 100 l p.e^{-1} d^{-1}, a 40% reduction of organic load in the primary treatment and an effluent quality requirement of below 40 mg/L BOD_5. This indicates that HSSF-CWs are area-requirement competitive when compared to the widely applied waste stabilization pond system in the tropics. Indeed a facultative pond achieving BOD removal efficiencies of 75-80 % in the tropics would require 2.0-5.0 m^2 p.e^{-1} (Kivaisi, 2001).

3.3.3   TSS removal

The effluent TSS concentrations were found to have a poor relationship (not shown) with the TSS loading rate, in agreement with observations in literature that outlet concentrations are not generally related to inlet concentrations (Kadlec and Wallace, 2009), as internal wetland processes, i.e. plant detritus material and microbial films present on media particles result in a TSS contribution and thus irreducible background TSS concentrations. However, the mean effluent concentrations from the planted cells (Table 3.2) achieved levels well below the recommended 35 mg/L maximum permissible limit for discharge to surface water courses in Kenya (NEMA-Kenya, 2003). The removal of TSS in HSSF-CWs is usually effective with most of the suspended solids filtered out and settled within the first few meters beyond the inlet zone (Vymazal, 2005). In this study, we used coarse gravel without fine particles. Suspended solids in HSSF-CW are mainly removed by physical mechanisms, such as filtration, interception and sedimentation processes (Kadlec and

Wallace, 2009). Manios et al. (2003) found that beds with gravel performed better concerning TSS removal than beds with soil, sand and compost. Incase of high loading rates and poor biodegradability, the entrapment of particulate matter within the filtration media in HSSF-CW may however lead to problems of hydraulic conductivity (Kadlec and Wallace, 2009; Knowles et al., 2011).

Reduction in TSS effluent concentration was significantly higher only for the effluent of the planted wetland cell Cyp2 compared to that of the unplanted cell (p< 0.007). While the mass TSS removal rates showed a strong positive linear relationship with TSS mass removal rates in the pilot cells up to some point (Fig. 3.5), the vegetated cells sustained the linear relationship better than the unplanted cell up to a TSS loading rate of about 20 g m^{-2} d^{-1}, suggesting a positive influence by the emergent and rooted *Cyperus papyrus* macrophyte. The main contribution of emergent macrophytes to TSS removal in HSSF-CW is through the growth of their roots and rhizomes, which stabilize the wetland bed and minimizes the resuspension of sediment particles, moreover, at constant hydraulic loads, the rhizosphere contributes to increased interception and sedimentation (Bojcevska and Tonderski, 2007; Karathanasis et al., 2003). Nevertheless, other studies have found no difference in TSS removal between planted and unplanted beds (Konnerup et al., 2009; Manios et al., 2003).

3.4   Nitrogen removal

Ammonia and nitrate are the most important inorganic forms of nitrogen in wetlands. The ionized form of ammonia (NH_4^+) is predominant in most wetland systems because of moderate pH and temperature (Kadlec and Wallace, 2009). The main mechanism by which treatment wetlands remove nitrogen from wastewater is identified by many authors to be microbial mediated sequential nitrification-denitrification and thus depends much on the environment inside the system (Trang et al., 2010; Vymazal and Kröpfelová., 2009). The nitrification process requires about 4.3 g of O_2 per g of ammonium nitrogen oxidized (Schäfer et al., 1998) and the availability of sufficient amounts of oxygen is often the limiting factor for ammonium removal in treatment wetlands. The mean ammonium removal or nitrification rates obtained in the study, ranging from 2.1 ± 0.43 g m^{-2} d^{-1} to 2.7 ± 1.9 g m^{-2} d^{-1} in the planted cells (Table 3.3) were probably to the maximum capacity for NH_4^+-N in the studied pilot cells, when compared with those found in the literature for nitrification rates in wetlands, ranging between 0.01–2.15 g N m^{-2} d^{-1} with the mean value of 0.048 g N m^{-2} d^{-1} (Mayo and Bigambo, 2005; Vymazal, 2007).

NO_3^--N concentration levels showed no significant difference in the influent and effluent concentrations, suggesting that all nitrate formed from nitrification of ammonia was consumed within the wetland, probably as an electron acceptor in anoxic respiration processes (denitrification) that results in reduction of nitrate to nitrogen gas (N_2). Evaluating the nitrification scenario assuming the decrease in ammonium concentration was solely due to oxidation, the corresponding amount of oxygen required for the observed nitrification rates in the planted cells is computed to be 7.2-19.8 g m^{-2} d^{-1}. However, the oxygen consumption is much higher than that which can possibly be transferred by the influent wastewater and *Cyperus papyrus* roots considering the reported oxygen release rates of only 0.017 g m^{-2} d^{-1} (Kansiime and Nalubenga, 1999). This suggests that other mechanisms such as adsorption, sedimentation, assimilation into microbial and plant biomass, were playing a substantial role in nitrogen removal in the pilot HSSF-CW. Further, it is plausible in the oxygen-limited situation of the pilot HSSF-CW cells that other microbial pathways for nitrogen removal were active i.e., partial-nitrification of ammonium to nitrite combined to anaerobic ammonium oxidation (ANAMMOX) are possible (Dong and Sun, 2007). Ammonia volatilization is not expected to have contributed to ammonia removal as the mechanism is only significant for ammonia removal when the pH exceeds 10 (Garcia et al., 2010), while in this study, the effluent pH of the HSSF-CW cells was circum-neutral (Table 3.2).

The potential rate of nutrient (ammonium and nitrate) uptake by plants is limited by their net productivity (biomass) and the concentration of nutrients in plant tissues, while nitrate uptake by wetland plants is presumed to be less favored than ammonium uptake (Kadlec and Wallace, 2009). The *Cyperus papyrus* macrophyte nutrient uptake value for nitrogen amounts to 0.135 g m^{-2} d^{-1} in natural wetlands (Muthuri et al., 1989), with a high content of nutrients observed in the aerial biomass of *Cyperus papyrus*, an indication of active translocation and storage of nutrients to sites where they are needed for primary growth, e.g. synthesis of amine acids and enzymes

(Kansiime et al., 2007; Kyambadde et al., 2005). In tropical regions, seasonal translocation activity is very low, and several harvests can be made during the year, so plant uptake could play a significant role in nitrogen removal, especially in lightly loaded systems (Vymazal, 2007).

3.5   Phosphorus removal

The media used for HSSF-CW (e.g. pea gravel, crushed stones) usually do not contain great quantities of Fe, Al or Ca for ligand exchange reactions with phosphate and the HSSF-CW are rarely operated at low enough phosphorous loading rates to allow for phosphorous removal processes. Therefore, removal of phosphorus is generally low (Vymazal, 2005). The low TP removal rates observed in this study (Table 3.3) are not different from what is reported in literature. As indicated in a worldwide experience of phophorus removal in HSSF-CW described in Vymazal (2005), phosphorus removal rates are rather low in CW systems: at an average mass removal rate of 0.12 g P m^{-2} d^{-1} (45 g P m^{-2} year $^{-1}$) and an average mass-based efficiency of 32%. Indeed, phosphorus removal is usually not a primary design consideration of most HSSF-CW that are rather designed for BOD_5 and TSS removal. In subsurface flow wetlands, soluble phosphorus will move with the water flow, while phosphorus associated with particulate matter will be influenced by filtration and interception mechanisms present in the wetland bed (Kadlec and Wallace, 2009).

The low PO_4^{3-}-P removal rates observed in this study (Table 3.3) are not different from what is reported in the literature. For example, Okurut (1999) determined an average o-PO_4^- removal rate of 0.05 g m^{-2} d^{-1} in a *Cyperus papyrus* constructed wetland in Uganda. Removal of the soluble reactive phosphorus (PO_4^{3-}-P) was only observed in the planted cells of the pilot HSSF-CW, indicating that plants play a role in the removal of phophorus (by uptake of soluble reactive phosphorus with conversion to tissue phosphorus). Muthuri et al. (1989) reported *Cyperus papyrus* phosphorus uptake values of   0.0192 g m^{-2} d^{-1} in the Lake Naivasha wetlands. Other studies also show that plants remove phosphorus from the wastewater through plant uptake, but the quantity is often small compared with the loading rates (Bojcevska and Tonderski, 2007; Garcia et al., 2010; Konnerup et al., 2009). The observed increase of

PO_4^{3-}-P concentration in the effluent of the unplanted and unshaded control cell may have been contributed by the death and decay of algae.

3.6   Conclusion

This study revealed successful performance of the tropical HSSF-CW for the secondary treatment of domestic wastewater with respect to organic matter (BOD_5 and COD) and TSS removal. For these parameters the effluent met the admissible local standards for discharge into surface water courses, at fairly short hydraulic retention times. The influence of *Cyperus papyrus* rhizosphere oxygen release may not have been sufficient to meet the oxygen demand of the wastewater. Indeed, anaerobic conditions prevailed in the HSSF-CW bed, as evidenced by high sulphate removal and moderate nitrification rates. The organic matter removal rates showed the HSSF-CW to be area requirement competitive, with a potential (at optimum hydraulic loading rates) of achieving high effluent quality for the secondary treatment of domestic wastewater under tropical conditions.

The tropical constructed wetland merits as a viable alternative to conventional treatment of domestic wastewater. Given the minimal maintenance requirements, the ease of operation and the good removal performance of bulk pollutants, the inexpensive constructed wetland technology can help to alleviate the current wastewater management problem in developing countries of discharging partially treated or untreated domestic wastewater into freshwater resources.

3.7   References

Abira, M.A., 2007. A pilot constructed treatment wetland for pulp and paper mill wastewater: performance, processes and implications for the Nzoia river, Kenya. PhD Thesis, UNESCO-IHE, Delft, Netherlands.

AMCOW, 2006. (African Ministers' Council on Water and Sanitation Program) Getting Africa on Track to Meet the MDGs on Water Supply and Sanitation - A Status Review of Sixteen African Countries, pp.33-43.

Baptista, J.D.C., Donnelly, T., Rayne, D., Davenport, R.J., 2003. Microbial mechanisms of carbon removal in subsurface flow wetlands. Water Science and Technology 48(5) 127-134.

Bojcevska, H., Tonderski, K., 2007. Impact of loads, season, and plant species on the performance of a tropical constructed wetland polishing effluent from sugar factory stabilization ponds. Ecological Engineering 29(1) 66-76.

Diemont, S.A.W., 2006. Mosquito larvae density and pollutant removal in tropical wetland treatment systems in Honduras. Environment International 32(3) 332-341.

Dong, Sun, T., 2007. A potential new process for improving nitrogen removal in constructed wetlands--Promoting coexistence of partial-nitrification and ANAMMOX. Ecological Engineering 31(2) 69-78.

Faulwetter, J.L., Gagnon, V., Sundberg, C., Chazarenc, F., Burr, M.D., Brisson, J., Camper, A.K., Stein, O.R., 2009. Microbial processes influencing performance of treatment wetlands: A review. Ecological Engineering 35(6) 987-1004.
Gagnon, V., Chazarenc, F., Comeau, Y., Brisson, J., 2007. Influence of macrophyte species on microbial density and activity in constructed wetlands. Water Science and Technology 56(3) 249-254.

Garcia, J., Aguirre, P., Mujeriego, R., Huang, Y.M., Ortiz, L., Bayona, J.M., 2004. Initial contaminant removal performance factors in horizontal flow reed beds used for treating urban wastewater. Water Research 38(7) 1669-1678.

Garcia, J., Rousseau, D.P.L., MoratÓ, J., Lesage, E.L.S., Matamoros, V., Bayona, J.M., 2010. Contaminant Removal Processes in Subsurface-Flow Constructed Wetlands: A Review. Critical Reviews in Environmental Science and Technology 40(7) 561-661.

Haberl, R., 1999. Constructed wetlands: A chance to solve wastewater problems in developing countries. Water Science and Technology 40(3) 11-17.

Jones, Muthuri, F.M., 1985. The Canopy Structure and Microclimate of Papyrus (Cyperus Papyrus) Swamps. Journal of Ecology 73(2) 481-491.

Kadlec, Reddy, K.R., 2001. Temperature Effects in Treatment Water Environment Research 73(5).
Kadlec, Wallace, S., 2009. Treatment wetlands. 2nd ed. Boca Raton, Fla: CRC Press, 1048 pp.

Kansiime, F., Nalubenga, M., 1999. Wastewater treatment by a natural wetland: the Nakivubo swamp Uganda -processes and implications. PhD Thesis, UNESCO-IHE, DELFT, Netherlands.

Kansiime, F., Saunders, M., Loiselle, S., 2007. Functioning and dynamics of wetland vegetation of Lake Victoria: an overview. Wetlands Ecology and Management 15(6) 443-451.

Karathanasis, A.D., Potter, C.L., Coyne, M.S., 2003. Vegetation effects on fecal bacteria, BOD, and suspended solid removal in constructed wetlands treating domestic wastewater. Ecological Engineering 20(2) 157-169.

Kibetu, P.M., 2008. Secondary and tertiary treatment of sewage in Kenya using constructed wetlands and gravel beds Proceedings of 2nd International Civil Engineering Conference on Civil Engineering and Sustainable Development, 23-25th September, Mombasa, Kenya.

Kivaisi, A.K., 2001. The potential for constructed wetlands for wastewater treatment and reuse in developing countries: a review. Ecological Engineering 16(4) 545-560.

Knowles, P., Dotro, G., Nivala, J., García, J., 2011. Clogging in subsurface-flow treatment wetlands: Occurrence and contributing factors. Ecological Engineering 37(2) 99-112.

Konnerup, D., Koottatep, T., Brix, H., 2009. Treatment of domestic wastewater in tropical, subsurface flow constructed wetlands planted with Canna and Heliconia. Ecological Engineering 35(2) 248-257.

Kyambadde, Kansiime, F., Dalhammar, G., 2005. Nitrogen and phosphorus removal in substrate-free pilot constructed wetlands with horizontal surface flow in Uganda. Water Air and Soil Pollution 165(1-4) 37-59.

Lee, C.-G., Fletcher, T.D., Sun, G., 2009. Nitrogen removal in constructed wetland systems. Engineering in Life Sciences 9(1) 11-22.

Manios, T., Stentiford, E.I., Millner, P., 2003. Removal of total suspended solids from wastewater in constructed horizontal flow subsurface wetlands. Journal of Environmental Science and Health - Part A Toxic/Hazardous Substances and Environmental Engineering 38(6) 1073-1085.

Mara, D., 2004. Domestic Wastewater Treatment in Developing Countries. Earthscan UK and US. Cromwell Press, Trowbridge, UK.

Mashauri, D.A., Mulungu, D.M.M., Abdulhussein, B.S., 2000. Constructed wetland at the University of Dar es Salaam. Water Research 34(4) 1135-1144.

Mayo, A.W., Bigambo, T., 2005. Nitrogen transformation in horizontal subsurface flow constructed wetlands I: Model development. Physics and Chemistry of the Earth 30(11-16) 658-667.

Mogaka, H., Gichere, S., Davis, R., Hirji, R., 2006. Climate Variability and Water Resources Degradation in Kenya. Improving Water Resources Development and Management. World bank working paper No. 69.

Muthuri, Jones, M.B., Imbamba, S.K., 1989. Primary productivity of papyrus (Cyperus papyrus) in a tropical swamp; Lake Naivasha, Kenya. Biomass 18(1) 1-14.

NEMA-Kenya, 2003. Environmental protection (Standards for effluent discharge) Regulations. General Notice No.44.of 2003.

Nyakang'o, J.B., van Bruggen, J.J.A., 1999. Combination of a well functioning constructed wetland with a pleasing landscape design in Nairobi, Kenya. Water Science and Technology 40(3) 249-256.

Nzengy'a, D.M., Wishitemi, B.E.L., 2001. The performance of constructed wetlands for, wastewater treatment: a case study of Splash wetland in Nairobi Kenya. Hydrological Processes 15(17) 3239-3247.

Okurut, T.O., 2000. A Pilot Study on Municipal Wastewater Treatment Using a Constructed Wetland in Uganda. PhD dissertation, UNESCO-IHE, Institute for Water Education, Delft, The Netherlands.

Pokorski, U., Onyango, P., 2010. Upscaling Access to Sustainable Sanitation. Follow-up Conference of the International Year of Sanitation (IYS), January 26, 2010,Tokyo, Japan, January 26, 2010.

Schäfer, D., Schäfer, W., Kinzelbach, W., 1998. Simulation of reactive processes related to biodegradation in aquifers: 1. Structure of the three-dimensional reactive transport model. Journal of Contaminant Hydrology 31(1-2) 167-186.

Tanner, 2001. Plants as ecosystem engineers in subsurface-flow treatment wetlands. Water Science and Technology 44(11-12) 9-17.

Tanner, D'Eugenio, J., McBride, G.B., Sukias, J.P.S., Thompson, K., 1999. Effect of water level fluctuation on nitrogen removal from constructed wetland mesocosms. Ecological Engineering 12(1-2) 67-92.

Trang, N.T.D., Konnerup, D., Schierup, H.-H., Chiem, N.H., Tuan, L.A., Brix, H., 2010. Kinetics of pollutant removal from domestic wastewater in a tropical horizontal subsurface flow constructed wetland system: Effects of hydraulic loading rate. Ecological Engineering 36(4) 527-535.

Tsagarakis, K.P., Mara, D.D., Angelakis, A.N., 2001. Wastewater management in Greece: experience and lessons for developing countries. Water Science and Technology 44(6) 163-172.

Verhoeven, J.T.A., Meuleman, A.F.M., 1999. Wetlands for wastewater treatment: Opportunities and limitations. Ecological Engineering 12(1-2) 5-12.

Vymazal, 1996. The use of subsurface-flow constructed wetlands for wastewater treatment in the Czech Republic. Ecological Engineering 7(1) 1-14.

Vymazal, 2005. Horizontal sub-surface flow and hybrid constructed wetlands systems for wastewater treatment. Ecological Engineering 25 478–490.

Vymazal, 2007. Removal of nutrients in various types of constructed wetlands. Science of the Total Environment 380(1-3) 48-65.

Vymazal, Kröpfelová, L., 2009. Removal of organics in constructed wetlands with horizontal sub-surface flow: A review of the field experience. Science of the Total Environment 407(13) 3911-3922.

Vymazal, Kröpfelová., 2009. Removal of nitrogen in constructed wetlands with horizontal sub-sureface flow: a review. Wetlands 29(4) 1114-1124.

Wanjogu, S.N., Kamoni, P.T., 1986. Soil conditions of Juja Estate (Kiambu district). Kenya Soil Survey project, Site evaluation Report no. P-79.

WB-WSP, 2012. Africa: Economics of Sanitation Initiative. Water and Sanitation Programme of the World Bank. http://www.wsp.org/wsp/content/africa-economic-impacts-sanitation. Worldbank.org.

Wiessner, A., Kappelmeyer, U., Kuschk, P., Kästner, M., 2005. Sulphate reduction and the removal of carbon and ammonia in a laboratory-scale constructed wetland. Water Research 39(19) 4643-4650.

Chapter 4: Performance comparison and economics analysis of waste stabilization ponds and horizontal subsurface flow constructed wetlands treating domestic wastewater

This chapter has been published as: Mburu, N., S.M. Tebitendwa, J.J.A. van Bruggen, D.P.L. Rousseau and P.N.L. Lens, 2013. Performance comparison and economics analysis of waste stabilization ponds and horizontal subsurface flow constructed wetlands treating domestic wastewater: A case study of the Juja sewage treatment works. Journal of Environmental Management, 128: 220-225.

Abstract

The performance, effluent quality, land area requirement, investment and operation costs of a full-scale waste stabilization pond (WSP) and a pilot scale horizontal subsurface flow constructed wetland (HSSF-CW) at Jomo Kenyatta University of Agriculture and Technology (JKUAT) were investigated between November 2010 to January 2011. Both systems gave comparable medium to high levels of organic matter and suspended solids removal. However, the WSP showed a better removal for Total Phosphorus (TP) and Ammonium (NH_4^+-N). Based on the population equivalent calculations, the land area requirement per person equivalent of the WSP system was 3 times the area that would be required for the HSSF-CW to treat the same amount of wastewater. The total annual cost estimates consisting of capital, operation and maintenance (O&M) costs were comparable for both systems. However, the evaluation of the capital cost of either system showed that it is largely influenced by the size of the population served, local cost of land and the construction materials involved. Hence, one can select either system in terms of treatment efficiency. When land is available other factor including the volume of wastewater or the investment, and O&M costs determine the technology selection.

4.0   Introduction

Conventional treatment plants are widely used for the treatment of domestic and industrial wastewater in developed countries. Even though very efficient, they are expensive to construct, intensive to maintain and require skilled personnel for their operation. In developing countries, very few well working conventional treatment plants can be found (Diaz and Barkdoll, 2006). Possible alternatives are those systems that provide a cheap, effective, reliable and sustainable way of treating wastewater. This includes the waste stabilization ponds (Babu, 2011) and constructed wetlands (Mburu et al., 2013). Both are well-established methods for wastewater treatment in tropical and subtropical climates (Machibya and Mwanuzi, 2006). Some of the advantages of these natural treatment systems over conventional wastewater treatment plants include: no need for skilled labour and therefore low operation and maintenance costs (Kayombo et al., 2000). Additionally, their zero-energy demand for the removal of organics and pathogenic organisms is contributing as a valuable tool for sustainable development (Thurston et al., 2001; Mashauri and Kayombo, 2002). It

66

is particularly the case when topography allows gravity feeding. However, one of the disadvantages of these natural systems is their space requirement (Bastos et al., 2010).

The main expenses related to sewage services are capital cost, operation and maintenance (O&M) costs and the procurement of land (Tsagarakis et al., 2003). These are thus important parameters for selecting an appropriate treatment system (Tsagarakis et al., 2003; Sato et al., 2007; Rousseau et al., 2008). In this sense, appropriate technology should be affordable (capital cost), have a low operation and maintenance cost (sustainability), be effective in meeting the discharge standards (efficiency), give the least nuisance (public acceptability) and be environmentally friendly (Mara, 2004; Mara et al., 2007). Thus natural wastewater treatment processes (i.e., non-electromechanical, using physical and biological processes) that are simple, low-cost and low-maintenance are preferred as appropriate alternatives for conventional wastewater treatment by any country, but especially in developing countries in the tropical areas (Mara, 2006; Sato et al., 2007).

Waste stabilization ponds (WSPs) and horizontal subsurface flow constructed wetlands (HSSF-CWs) are not equally applied in hot climates, as the latter technology has been recently embraced in the developing countries (Kivaisi, 2001; Mara, 2004; Li, 2007). Comparing land area requirements and costs to achieve a required effluent quality can be useful in deciding whether to use a HSSF-CW or a WSP (Mara, 2006). The aim of this chapter is to compare the two technologies in terms of pollutant removal efficiency, land area requirement, as well as investment, operation and maintenance costs. Such a comparison based on available data and facilities could offer technical and economic insights that would simplify technology selection processes. This deserves consideration and is especially relevant to small communities where the need for a reliable treatment of wastewater not only integrates with the selection of an economically and environmentally viable wastewater treatment technology, but also with social attributes such as workforce education level and available land. In this study, domestic wastewater was send to a pre-existing treatment system consisting of three interconnected waste stabilization ponds- a primary facultative and secondary facultative pond, followed by a maturation pond. An experimental pilot scale HSSF-CW was introduced and fed from the outflow of the primary facultative pond.

4.1 Experimental set-up

4.1.1 Study area

The pilot scale HSSF-CW and the WSP were located within the premises of the Jomo Kenyatta University of Agriculture and Technology (JKUAT) sewage works, in Juja town, Kenya. The sewage works are at a global position of 1°05'49.28"S 37°01'21.91"E and an altitude of 1463 m, some 40 km North East of Nairobi City. JKUAT sewage works treats the university campus' domestic wastewater of about 5000 inhabitants by a set of 5 stabilization ponds. The final effluent is discharged to a near-by river via a natural waterway (Fig.4.1).

4.1.2 Description of the systems

The system described has an operational WSP and an experimental pilot scale HSSF-CW. The WSP system designed for a 2700 population equivalent, consisted of 2 primary facultative ponds (PFP), 2 secondary facultative ponds (SFP) and a maturation pond (MP) that are aligned in series (Fig. 4.1). The primary facultative ponds also act as the sedimentation pond receiving raw wastewater from the septic tank effluent pump. The septic tank (located 600 m away from the waste stabilization ponds) receives wastewater from the students' hostels, staff quarters, cafeterias and laboratories at JKUAT. The ponds were periodically covered with duckweed (*Lemna sp.*) and Fern (*Azolla pinnata*) and attracted ducks and goose. The mean flow rate into the PFP and SFP was 349 m³/d and that into the MP was 698 m³/d, resulting into a hydraulic retention time (HRT) of 14.6, 6.9, and 19.5 days respectively, giving a total HRT of 41 days.

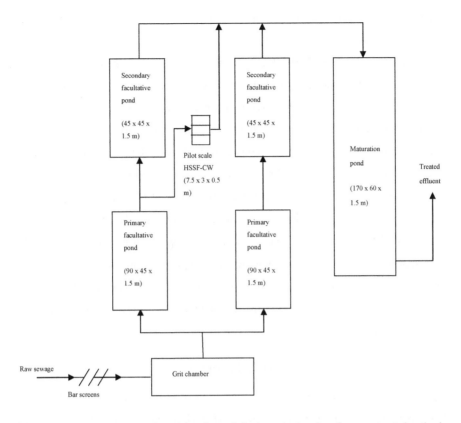

Fig. 4.1. Waste stabilization ponds and the pilot scale horizontal subsurface flow constructed wetlands schematic layout at the JKUAT sewage treatment works

The constructed wetland system, described in details by Mburu et al. (2013), consisted of two pilot scale cells of HSSF-CW receiving primary effluent from the outflow of one of the PFP. In one of the HSSF-CW cells, the macrophyte *Cyperus papyrus* was planted (1 plant per m^2), while the second cell acted as a control. At the time of this study, *C. papyrus* had senescenced, but below-ground biomass was present. The mean flow rate for the control and planted cell was 3 m^3/d, resulting in a HRT of 1.5 days.

In WSP holding basins or lagoons, wastewater is stabilized in a confined environment. Area requirements are based on volumetric and surface organic loading rates (Mara, 2006). In the HSSF-CW, water flows through a porous medium such as gravel in which plants are rooted. The porous media provide anchorage for the emergent macrophyte and surface area for microbial growth which consequently enhances the transformation or removal of chemical and biological constituents in wastewater (García et al., 2010). Determination of area requirements for constructed wetlands is based on the rules of thumb or first order design equations (Kadlec and Wallace, 2009). The wastewater has to be pretreated prior to the HSSF-CW (a minimal acceptable level being equivalent to primary treatment), to reduce the risk of clogging from solids. The biological conditions in these systems are similar in certain aspects; water near the bottom is in an anoxic/anaerobic state, while a shallow zone near the water surface tends to be aerobic. The source of oxygen is atmospheric re-aeration, photosynthesis (in facultative pond) and root oxygen leaching (in wetlands). These natural treatment systems promote the development of microorganisms which are responsible for much of the biological treatment occurring in the systems. The operation of the two systems does not involve material input (chemicals) and energy consumption, as treatment mechanisms are natural and wastewater flow within the systems is entirely driven by gravity.

4.1.3   Analysis of the physico-chemical parameters

Physico-chemical variables were measured *in-situ*. pH, conductivity and total dissolved solids were measured with a HACH, ECO 40 multi-probe, whereas dissolved oxygen and temperature were measured with a dissolved oxygen meter (Model HACH, HQ40d multi probe).

Wastewater samples were collected for analysis of ammonium-nitrogen (NH_4^+-N), total phosphorus (TP), chemical oxygen demand (COD), biological oxygen demand (BOD_5) and total suspended solids (TSS). All were analyzed using standard methods as described in APHA (1998). Mass removal efficiencies were calculated as a difference between input and output mass loading rates. Statistical analysis was performed using MINITAB 15 and Microsoft Excel 2007 software. Comparison of variables was performed using the analysis of variance technique (ANOVA) and the Tukey's multiple comparisons was used to test differences in the means.

4.1.4 Cost evaluation

The costs of the two systems were calculated based on land requirement, capital, operation and maintenance costs for the treatment of domestic wastewater at JKUAT sewage works, and expressed per population equivalent (PE). Comparative costing was based on undiscounted (and annualized) capital, O&M cost estimates at the time of the study, and ignoring the need for expenditures to earn a rate of return on the investment. Capital cost included the cost of land and the estimated construction cost for both the WSP (earthworks, lining the bottom and embankments, inlet and outlet structures) and the HSSF-CW (masonry and concrete works, plumbing, gravel media and establishing the macrophyte). O&M costs included those of personnel and maintenance (routine and arising repairs, inlet-outlet inspection, landscaping, desludging and macrophyte harvesting (for the HSSF-CW)). A design period of 20 years for the HSSF-CW and WSP systems was used. The initial cost calculation for the HSSF-CW is based on the pilot plant sizing of 12 PE, which is eventually extrapolated to a full size HSSF-CW for 2700 PE and with a corresponding area of the sedimentation pond included in the costing as well.

4.2   Results and discusion

4.2.1   Quality of influent and effluent

The main results for the water quality parameters (BOD$_5$, COD, TSS, NH$_4^+$-N, and TP) are given in Table 4.1 for the PFP, SFP, MP and the HSSF-CWs. The WSP removed almost or more than 90 % of the BOD, COD and TSS, only 21 % of the TP and 50 % of the NH$_4^+$-N. The HSSF-CW removed more than 84 % of BOD$_5$, COD and TSS. NH$_4^+$-N was not removed while, TP removal was 13 % in the unplanted (control) and 26 % in the planted HSSF-CW.

The WSP effluent values are within the Kenya National Environmental Management Authority (NEMA) effluent discharge guidelines, except for NH$_4^+$-N and TP (NEMA-Kenya, 2003). The planted and the unplanted HSSF-CWs exceeded the discharge limit values for NH$_4^+$-N and TP as well. For the planted HSSF-CW, it must be noted that the plants (above ground biomass) were harvested and only below-ground plant parts, i.e. live *Cyperus papyrus* roots and rhizomes, were present in the HSSF-CW during the study. Compared with the planted HSSF-CW, effluent values for pollutants

71

from the unplanted HSSF-CW are relatively high (Table 4.1). This is an indication that plants play a role in the wastewater purification processes of the HSSF-CW (Kadlec and Knight, 1996; Stottmeister et al., 2003).

Table 4.1. Effluent water quality values and removal efficiencies of the different systems

System	BOD_5			COD			TSS			NH_4^+N			TP		
	mg/l	SD	%	mg/l	SD	%	mg/l	SD	%	mg/l	SD	%	mg/l	SD	%
Influent	232	133.3		424	277.4		118	87.6		39	13.6		4	0.9	
PFP	74	17.7	68	216	110.9	49	56	14.9	52	34	10.6	14	4	0.7	5
SFP	48	14.6	79	160	77.3	62	39	22.9	67	35	13.8	11	3	0.6	21
MP	20	16	91	100	37	76	10	20	91	**17**	11	56	**3**	0.6	21
HSSF-CW (Planted)	29	9	87	58	17.4	86	19	8.7	84	**36**	8	8	**3**	0.3	26
HSSF-CW (Unplanted)	39	14.5	83	98	17.6	77	34	15	71	**39**	5.6	0	**3**	0.6	13

Values in bold don't comply with the NEMA standards. (NEMA guidelines for effluent discharge into surface waters are: less than 120, 40, 35, 1 and 1 mg l for, COD, BOD5, TSS and ammonium compound, TP, respectively (NEMA 2003)).SD: Standard deviation

It also should be noted that in this set-up the HSSF-CW are providing up to secondary treatment only. Nevertheless the HSSF-CW achieved a better effluent quality with regard to the bulk pollutants, i.e. organic matter and the suspended solids compared to the WSP (Table 4.1). This is attributed to the difference in the type of treatment (i.e. biofilm attached to the gravel media in the HSSF-CW and suspended growth in the WSP) and the generation of secondary BOD and TSS from algal and duckweed biomass in the WSP. Thus despite the low HRT of only 1.5 days in the HSSF-CW, compared to the recommended up to 8 days for acceptable removal efficiency of organic matter (Akratos and Tsihrintzis, 2007), the HSSF-CW was found to provide a comparable treatment performance with the WSP. The WSP included an extra treatment (tertiary) step and an overall longer retention time.

The higher NH_4^+-N concentration (Table 4.1) and the resulting poor removal performance observed in the HSSF-CW compared to the WSP effluent are attributed to the inadequate re-oxygenation capacity in the subsurface flow environment of the HSSF-CW system (Okurut, 2000; Mburu et al., 2012). Indeed, the direct contribution by atmospheric re-aeration and algal photosynthesis, as is the case for a WSP is insignificant in the global aeration process of the HSSF-CW. This reduces the extent of possible nitrification and subsequently limits denitrification due to a lack of nitrates. This was clear from the consistent drop in dissolved oxygen (DO) between the influent and effluent, with a mean effluent DO concentration of 0.6 mg/l, in contrast to an increase to 3.2 mg/l DO at the WSP effluent (data not shown).

Kadlec and Knight (1996) further stress that a HSSF-CW system is good in nitrate removal (denitrification), but not for ammonia oxidation since oxygen availability is the limiting step in nitrification. On the other hand, similar poor ammonia removal was observed for the secondary facultative pond (SFP) despite the high mean oxygen concentration in its effluent (mean 10.2 mg/l) compared to the influent wastewater (1.3 mg/l). There was only a slight decrease in ammonium concentration in the SFP, which was finally reduced to 17 mg/l in the MP effluent. The results, nonetheless, are comparable to the findings of Babu (2011), who reported an increase in the ammonium concentration in the facultative pond effluent and a reduction in the ammonium concentration in the MP effluent. The mineralization of organic compounds within the aerobic layer of the SFP due to the supply of oxygen from algae could be responsible for the high NH_4^+-N concentration observed in the SFP effluent (Senzia et al., 2003; Shammas et al., 2009). Further, lack of attachment surfaces for the nitrifiers could be responsible for the high NH_4^+-N concentration in the effluent of WSP (Zimmo et al., 2003; Babu, 2011). Several studies have shown that the introduction of attachment surface for nitrifiers in the ponds improves nitrogen removal (Pearson, 2005; Babu, 2011).

Phophorus removal through assimilation by algae to support cell synthesis seems to have been considerable in the WSP judging from the higher removal rate calculated at 2.54 g TP $m^{-2}d^{-1}$ compared to 0.13 g TP $m^{-2}d^{-1}$ achieved in the HSSF-CW. Reasons for the lower phophorus removal efficiency of the HSSF-CW system may be related to the lower sorption and retention capacity of the granitic gravel media and the saturation of phosphorus uptake by plants.

4.2.2 Performance of the systems based on area and HRT

The WSP receives the wastewater of 2700 PE and has a total area of 22350 m^2 and a hydraulic loading rate of 698 m^3/day, resulting in a HRT of 41 days, and an area of 8.3 m^2 per PE. Since the HSSF-CW is receiving the influent from the PFP, it does not need a sedimentation pond in this setup, and therefore the area per PE is rather small at 1.9 m^2. The sedimentation pond is of utmost importance to avoid clogging the HSSF-CW filter bed which results directly on the accumulation of suspended solids introduced with the influent (solids entrapment and sedimentation). In the calculations for cost, we assumed the need of a sedimentation basin, and therefore the area per PE increased from 1.9 to 3.4 m^2/PE. This area for the HSSF-CW is somewhat low compared to the applied hydraulic loading and to values in the literature for secondary wastewater treatment with HSSF-CW estimated at 6 m^2/PE (Mara, 2006). This can be part of the explanation of the rather fair performance by the HSSF-CW. However, as the overall performance for both systems is comparable, it can be deduced that the area requirement for the HSSF-CW is 2.5 times less that of the WSP.

4.2.3 Cost evaluation

The cost for construction and maintenance of the WSP and the HSSF-CW at JKUAT have been calculated and Table 4.2 gives a detailed overview. The costs are based on a lifetime of 20 years and 2700 PE. Initially, the cost for the small pilot HSSF-CW (12 PE) is given, for which an extrapolation is made to match the 2700 PE to enable direct comparison with the WSP costs. The operation and maintenance tasks are supposed to be necessary in order to achieve optimal treatment and attain the expected durable and reliable performance (Rousseau et al., 2008). For example after desludging, these systems are ready to perform a new cycle of operation without having to change any electro-mechanical equipment. This ensures a lower operation cost as well as a reliability of performance which is not negatively affected by the occasional malfunctioning of such equipment.

Ninety-eight % of the construction cost of the WSP goes to the cost of land and the lining. Therefore, costs will be less when land is cheap and no liner is needed (e.g. due to soil conditions). For maintenance of the WSP the main costs are personnel and desludging. For the HSSF-CW, 93 % of the construction cost is in the purchase of land, the masonry and the concrete works. Thus land is an important cost factor, while a construction not involving concrete and masonry structures can save upto 40 % in construction cost. Indeed for both systems, the variability of the soil conditions at various different sites and the need to prevent leakage of the wastewater into the ground water will influence the cost that goes into the preparation of the sides and bottom. Again desludging is costly in the maintenance cycle of HSSF-CW accounting for 40 % of the operation and maintenance cost. The construction costs of the HSSF-CW are lower than those of the construction of the WSP, but the annual maintenance costs are higher. The overall costs per PE for the two systems are in the same range (Table 4.3). For the HSSF-CW it should be noted that both the area and costs have been extrapolated from the pilot scale system to a potential full scale systems, and in reality the total cost will be less considering the extensive concrete and masonry works in the experimental pilot scale HSSF-CW. Nevertheless, the costs for some tasks such as plant harvesting cannot be easily reduced because the type of equipment necessary is the same and this operation has to be done manually.

The determined land area requirement for the HSSF-CW to serve 2700 PE is three times less compared to that for the WSP. Such a footprint is especially an advantage in countries with high land prices. These results corroborate with those of Senzia *et al.* (2003), who also reported that a WSP needed more land area than a HSSF-CW based on the same effluent quality. However, Mara (2006) computed a higher land area requirement for a secondary HSSF-CW compared to a secondary facultative pond. The comparisons were made at three levels of effluent quality in the UK i.e. (i) that specified in the Urban Waste Water treatment Directive (UWWTD) of 25 mg filtered BOD l^{-1} and 150 mg suspended solids (SS) l^{-1} for WSP effluents, and 25 mg unfiltered BOD l^{-1} for CW (and all other) effluents, (ii) two common requirements of the Environment Agency (the environmental regulator for England and Wales) for small works (a) the "40/60" requirement i.e. 40 mg BOD l^{-1} and 60 mg SS l^{-1} and (b) the "10/15/5" requirement i.e. 10 mg BOD l^{-1}, 15 mg SS l^{-1} and 5 mg ammonia-N l^{-1}. Yet, it is difficult for WSP systems to meet such requirements, especially those for SS, so WSP effluents must be 'polished' (Johnson et al., 2007). In Brazil, Bastos et al. (2010) indicated that HSSF-CW and WSP require similar land areas to achieve a bacteriological effluent quality suitable for unrestricted irrigation (10^3 *E. coli* per 100 mL), but HSSF-CW would require 2.6 times more land area than ponds to achieve quite lenient ammonia effluent discharge requirement (20mg NH_3 L^{-1}), and, by far, more land than WSP to produce an effluent complying with the WHO helminth guideline for agricultural use (≤ 1 egg per liter).

Table 4.3 compares the area requirement and cost of the WSP and HSSF-CW. The investment cost for the WSP is 3 times more than for the HSSF-CW. However, the maintenance cost for the WSP is much less, making the total cost about equal.

Expressed per PE, the cost is about 13 euro per PE per year for maintenance and investment (based on a 20 years lifetime).

Table 4.2. Capital, operation and maintenance costs (EUROs) of WSP and HSSF-CW for 2700 PE at JKUAT (design lifetime of 20 years)

WSP (based on 2700 PE)

Capital/ investment cost	Unit	Rates	Quantity	Amount	%
Land cost	m^2	15	22350	335250	47.6
Excavation	m^3	0.5	28600	14300	2.0
Lining-sides+Bottom	m^2	5	17640	352800	50.1
Inlet-outlet structure	No.	250	10	2500	0.4
Total investment cost (Euros)				**704850**	
Annual Recurrent Cost					
Personnel	Manpower	86.4	2	172.8	61.1
Repair works	piecework	10	1	10	3.5
Desludging every 5 years	piecework	500	0.2	100	35.4
Total annual recurrent (Euros)				**282.8**	

CW (based on 11.6 PE)

Capital/ investment cost					
Land cost	m^2	15	39.9	598.5	50.6
Excavation	m^2	0.5	40	20	1.7
Masonry and Concrete works	No. (Piecework)	500	1	500	42.2
Cost of gravel	t	10	6	60	5.1
Planting	Piecework	5	1	5	0.4
Total investment cost (Euros)				**1183.5**	
Annual Maintenance Cost					
Maintenance (hired labour)	Piecework	5	6	30	30.0
Repairworks	Piecework	10	2	20	20.0
Plant harvesting	Piecework	10	1	10	10.0
Desludging every 5 years	piecework	200	0.2	40	40.0
Total cost (Euros)				**100**	

Table 4.3. Comparison of costs (Euros) and area (m^2) of the WSP and HSSF-CW based on population equivalent (PE)

Area/ Cost	System	
	WSP	**HSSF-CW**
Area	22350	39.9
Design PE	2700	11.6
Area per PE	8.3	3.4
Area for 2700 PE	22350	9287.1
Total Investment (2700 PE)	704850	275872
Investment/ yr (2700 PE)	35242	13747
Maintenance/yr (2700 PE)	283	23300
Total cost/yr (2700 PE)	35525	37047
Total cost/PE.yr	13.2	13.7

Table 4.4 compares the JKUAT results with Ahmed and van Bruggen (2010), where a WSP, various constructed wetland combinations and a floating wetland system were compared. Table 4.4 also includes data from a HSSF-CW and WSP in Uganda and an activated sludge (AS) system in The Netherlands (personal communication). The total cost of the "green" systems (WSP and CW) ranges from 12 - 33 Euro/PE.yr, with an average of 14.4 Euro/PE.yr, whereas the cost for the activated sludge wastewater treatment (AS) system is at least three times higher. Their area needed per PE ranges from 1.5 to 10.1 m^2/PE compared to 0.4 m^2/PE for the AS.

Although the "green" systems are not capable of achieving the same consistent treatment performance efficiency as activated sludge plants, many studies report quite acceptable levels of effluent quality of both WSPs and HSSF-CWs (Kadlec and Wallace, 2009; Vymazal, 2011). These good performances are also indicated by the data presented in this work (Table 4.1). However, what the decision makers are most interested in is the reliability of performance. For both the WSPs and HSSF-CWs, this is not as dependent on complex and strict operational constraints as it is for many conventional/intensive processes. Once they are properly designed, built, operated and maintained, the probability of malfunctioning of WSP and CWs is low (Rousseau et al., 2008; Kadlec and Wallace, 2009; Ahmad and Van Bruggen, 2010). Further, these technologies are quite suitable for the tropical and subtropical regions where temperature is normally high (Yu et al., 1997). It can be expected that biofilm formation in these regions will be higher and probably provide diverse microenvironments for effective wastewater treatment (Babu, 2011). As such, the systems can achieve the required effluent quality at treatment works serving either large populations or small communities ($\leq$ 2,000 PE).

Table 4.4. Comparison of the WSP and CW with various systems

System	HRT	PE	Area/PE	Euro/PE.yr	Description
WSP	41	2700	8.3	13.2	Waste stabilisation pond (This work)
HSSF-CW	1.5	12	3.4	13.7	Planted constructed wetland (This work)
CW1	24.3	105	5.3	19.4	Constructed wetland combination [a]
CW2	38.3	53	10.1	33.3	Constructed wetland combination [a]
CW3	21.4	97	5.1	20.9	Constructed wetland combination [a]
F-W	10.6	13	1.5	11.8	Floating wetland system [a]
WSP	73.4	65	9.4	12.8	Waste stabilisation pond [a]
CW-Uganda			1.8	2	CW without liner and concrete [a]
WSP-Uganda			2.1	2.5	Waste stabilization ponds without liner [b]
AS Netherlands			0.4	50	Activated sludge wastewater treatment plant [*]

[a]: Ahmed and van Bruggen (2010)   [b]: Okurut (2000)   [*]: personal communication

4.3   Conclusion

Comparable medium to high levels of organic matter and suspended solids removal were attained in both the WSP and HSSF-CW systems investigated. However, the WSP attained a better removal for NH_4^+-N.   The total annual cost estimates consisting of capital, operation and maintenance costs had little difference between both systems. However, the evaluation of the capital cost of either system showed that it is largely influenced by the cost of land and the required construction materials. The HSSF-CW showed less land requirement per unit volume of treated wastewater compared to that of the WSP.   Hence, one can select either system in terms of treatment efficiency. When land is abundantly available, other factor including the volume of wastewater to be treated and the economies of scale, determine the final costs.

4.4   References

Ahmad, A. and J.J.A. Van Bruggen, 2010. Purification efficiency and economics of hybrid constructed wetlands, floating wetlands and stabilization pond systems in a parallel treatment of domestic wastewater in Spain. Proceedings of the 12th IWA International Conference on wetland systems for water pollution control. 4-9 October, 2010, Venice, Italy. 6p.

Akratos, C.S. and V.A. Tsihrintzis, 2007. Effect of temperature, HRT, vegetation and porous media on removal efficiency of pilot-scale horizontal subsurface flow constructed wetlands. Ecological Engineering, 29: 173-191.

Babu, M., 2011. Effect of Algal Biofilm and Operational Conditions on Nitrogen Removal in Wastewater Stabilization Ponds. PhD dissertation, UNESCO-IHE, Institute for Water Education, Delft, The Netherlands.

Bastos, R.K.X., M.L. Calijuri, P.D. Bevilacqua, E.N. Rios, E.H.O. Dias, B.C. Capelete and T.B. Magalha˜ es, 2010. Post-treatment of UASB reactor effluent in waste stabilization ponds and in horizontal flow constructed wetlands: a comparative study in pilot scale in Southeast Brazil. WST, 61.

Diaz, J. and B. Barkdoll, 2006. Comparison of Wastewater Treatment in Developed and Developing Countries. World Environmental and Water Resource Congress 2006, pp. 1-10.

García, J., D.P.L. Rousseau, J. Morató, E.L.S. Lesage, V. Matamoros and J.M. Bayona, 2010. Contaminant Removal Processes in Subsurface-Flow Constructed Wetlands: A Review. Critical Reviews in Environmental Science and Technology, 40: 561-661.

Johnson, M., M.A. Camargo Valero and D.D. Mara, 2007. Maturation ponds, rock filters and reedbeds in the UK: statistical analysis of winter performance. Water Science & Technology 55 135–142.

Kadlec and R.L. Knight, 1996. Treatment wetlands Boca Raton, Fla.: CRC Press, 893 pp.

Kadlec and S. Wallace, 2009. Treatment wetlands. 2nd ed. Boca Raton, Fla: CRC Press, 1048 pp.

Kayombo, S., T.S.A. Mbwette, A.W. Mayo, J.H.Y. Katima and S.E. Jorgensen, 2000. Modelling diurnal variation of dissolved oxygen in waste stabilization ponds. Ecological Modelling, 127: 21-31.

Kivaisi, A.K., 2001. The potential for constructed wetlands for wastewater treatment and reuse in developing countries: a review. Ecological Engineering, 16: 545-560.

Li, L., Li, Y., Biswas, D.K., Nian, Y., Jiang, G., 2007. Potential of constructed wetlands in treating the eutrophic water: evidence from Taihu Lake of China. Bioresour Technol., 99: 1656-1663.

Machibya, M. and F. Mwanuzi, 2006. Effect of low quality effluent from wastewater stabilization ponds to receiving bodies, case of Kilombero sugar ponds and Ruaha river, Tanzania. Int J Environ Res Public Health, 3: 209-216.

Mara, D., 2004. Domestic Wastewater Treatment in Developing Countries. Earthscan UK and US. Cromwell Press, Trowbridge, UK.

Mara, D., 2006. Constructed wetlands and waste stabilization ponds for small rural communities in the United Kingdom: A comparison of land area requirements, performance and costs. Environmental Technology, 27: 753-757.

Mara, D., J.-O. Drangert, N.V. Anh, A. Tonderski, H. Gulyas and K. Tonderski, 2007. Selection of sustainable sanitation arrangements. Water Policy, 9: 305-318.

Mashauri, D.A. and S. Kayombo, 2002. Application of the two coupled models for water quality management: facultative pond cum constructed wetland models. Physics and Chemistry of the Earth, Parts A/B/C, 27: 773-781.

Mburu, N., D. Sanchez-Ramos, Diederik P.L. Rousseau, J.J.A van Bruggen, George Thumbi, Otto R. Stein, Paul B. Hook and P. N.L.Lens, 2012. Simulation of carbon, nitrogen and sulphur conversion in batch-operated experimental wetland mesocosms. Ecological Engineering.

Mburu, N., S. Tebitendwa, D. Rousseau, J. van Bruggen and P. Lens, 2013. Performance Evaluation of Horizontal Subsurface Flow–Constructed Wetlands for the Treatment of Domestic Wastewater in the Tropics. Journal of Environmental Engineering, 139: 358-367.

NEMA-Kenya, 2003. Environmental protection (Standards for effluent discharge) Regulations. General Notice No.44.of 2003.

Okurut, T.O., 2000. A Pilot Study on Municipal Wastewater Treatment Using a Constructed Wetland in Uganda. PhD dissertation, UNESCO-IHE, Institute for Water Education, Delft, The Netherlands.

Pearson, H., 2005. Microbiology of waste stabilization ponds. In: Pond Treatment Technology. A. Shilton (Ed). IWA publishing, London: 4-48.

Rousseau, D.P.L., E. Lesage, A. Story, P.A. Vanrolleghem and N. De Pauw, 2008. Constructed wetlands for water reclamation. Desalination, 218: 181-189.

Sato, N., T. Okubo, T. Onodera, L.K. Agrawal, A. Ohashi and H. Harada, 2007. Economic evaluation of sewage treatment processes in India. Journal of Environmental Management, 84: 447-460.

Senzia, M.A., D.A. Mashauri and A.W. Mayo, 2003. Suitability of constructed wetlands and waste stabilisation ponds in wastewater treatment: nitrogen transformation and removal. Physics and Chemistry of the Earth, Parts A/B/C, 28: 1117-1124.

Shammas, N., L. Wang and Z. Wu, 2009. Waste Stabilization Ponds and Lagoons. In: L. Wang, N. Pereira and Y.-T. Hung (Eds), Biological Treatment Processes, Vol. 8, Humana Press, pp. 315-370.

Stottmeister, U., A. Wießner, P. Kuschk, U. Kappelmeyer, M. Kästner, O. Bederski, R.A. Müller and H. Moormann, 2003. Effects of plants and microorganisms in constructed wetlands for wastewater treatment. Biotechnology Advances, 22: 93-117.

Thurston, J.A., K.E. Foster, M.M. Karpiscak and C.P. Gerba, 2001. Fate of indicator microorganisms, giardia and cryptosporidium in subsurface flow constructed wetlands. Water Research, 35: 1547-1551.

Tsagarakis, K.P., D.D. Mara and A.N. Angelakis, 2003. Application of Cost Criteria for Selection of Municipal Wastewater Treatment Systems. Water, Air, & Soil Pollution, 142: 187-210.

Vymazal, J., 2011. Long-term performance of constructed wetlands with horizontal sub-surface flow: Ten case studies from the Czech Republic. Ecological Engineering, 37: 54-63.

Yu, H., J.-H. Tay and F. Wilson, 1997. A sustainable municipal wastewater treatment process for tropical and subtropical regions in developing countries. Water Science and Technology, 35: 191-198.

Zimmo, O.R., N.P. van der Steen and H.J. Gijzen, 2003. Comparison of ammonia volatilisation rates in algae and duckweed-based waste stabilisation ponds treating domestic wastewater. Water Research, 37: 4587-4594.

Chapter 5:   Reactive transport simulation in a tropical horizontal subsurface flow constructed wetland treating domestic wastewater

This chapter has been published as: Mburu, N., Rousseau, D.P.L., van Bruggen, J.J.A., Thumbi, G., Llorens, E., García, J., Lens, P.N.L., 2013. Reactive transport simulation in a tropical horizontal subsurface flow constructed wetland treating domestic wastewater. Science of the Total Environment 449(0) 309-319.

Abstract

A promising approach to the simulation of flow and conversions in the complex environment of horizontal subsurface flow constructed wetlands (HSSF-CWs) is the use of reactive transport models, in which the transport equation is solved together with microbial growth and mass-balance equations for substrate transformation and degradation. In this study, a tropical pilot scale HSSF-CW is simulated in the recently developed CWM1-RETRASO mechanistic model. The model predicts organic matter, nitrogen and sulphur effluent concentrations and their reaction rates within the HSSF-CW. Simulations demonstrated that these reactions took place simultaneously in the same (fermentation, methanogenesis and sulphate reduction) or at different (aerobic, anoxic and anaerobic) locations. Anaerobic reactions occurred over large areas of the simulated HSSF-CW and contributed (on average) to the majority (68%) of the COD removal, compared to aerobic (38%) and anoxic (1%) reactions. To understand the effort and compare computing resources needed for the application of a mechanistic model, the CWM1-RETRASO simulation is compared to a process based, semi-mechanistic model, run with the same data. CWM1-RETRASO demonstrated the interaction of components within the wetland in a better way, i.e. concentrations of microbial functional groups, their competition for substrates and the formation of intermediary products within the wetland. The CWM1-RETRASO model is thus suitable for simulations aimed at a better understanding of the CW system transformation and degradation processes. However, the model does not support biofilm-based modeling, and it is expensive in computing and time resources required to perform the simulations.

5.0   Introduction

The constructed wetland technology has become useful in mitigating environmental pollution by taking advantage of natural processes for wastewater treatment (Hench et al., 2003; Kivaisi, 2001). However, constructed wetlands (CWs) exhibit a high complexity in their pollutant removal processes and mechanisms (Faulwetter et al., 2009; García et al., 2010; Ojeda et al., 2008; Reddy and D'Angelo, 1997; Wu et al., 2011). Lately, mechanistic models for CWs have become promising tools for a better description and an improved understanding of constructed wetland treatment processes and performance (Langergraber et al., 2009a; Moutsopoulos et al., 2011).

83

Their application involves the simulation of reactions and flow in the heterogeneous environment of subsurface flow CWs by the use of the advective–dispersive transport equations solved together with the microbial growth and mass-balance reaction equations for substrate transformation and degradation, i.e. the biokinetic models. A few CW biokinetic models have been formulated with different levels of complexity in terms of their mathematical structure, i.e. constitutive laws and equations, number of variables or components and parameters, etc. (Langergraber and Šimůnek, 2012). An example of these mechanistic models is the Constructed Wetland Model N°1 (CWM1), which main aim is to provide a widely accepted model formulation describing biochemical transformation and degradation processes for organic matter, nitrogen and sulphur in subsurface flow CWs (Langergraber et al., 2009c).

Mechanistic numerical models can be used to determine the relationships between the different wetland wastewater treatment processes and weigh their relative contributions (Ojeda et al., 2008). Compared to the empirical models and first order models, the mechanistic models for constructed wetlands are more versatile and accurate predictive tools over a much larger range of operating conditions (Bezbaruah and Zhang, 2004; Langergraber, 2007; Rousseau et al., 2004). Together with being resilient to perturbations, the mechanistic models offer the opportunity of testing the sensitivities of the pollutant removal processes in different operational conditions when the evaluated scenarios cannot be easily tested physically (Langergraber, 2011; Langergraber and Šimůnek, 2012; Liolios et al., 2012; Llorens et al., 2011b; Min et al., 2011; Rousseau, 2005).

In order to simulate the reactive transport and treatment performance in CWs, the biokinetic models have been coupled to hydrodynamic models (Langergraber et al., 2009b) and implemented in software to solve the differential equations for dynamic simulations (Mburu et al., 2012). One recent work implemented the CWM1 processes within the finite element code RetrasoCodeBright (RCB), obtaining the CWM1-RETRASO model (Llorens et al., 2011a). This 2D mechanistic model simulates hydraulics and reactive transport as well as the main microbial reactions for organic matter, nitrogen and sulphur biodegradation and transformation in horizontal subsurface flow constructed wetlands (HSSF-CWs) (Llorens et al., 2011b).

Generating and fitting data for a diverse and wide range of pollutant loading rates and hydraulic retention times will illustrate the robustness and predictive power of such mechanistic models for CWs. In this work, the CWM1-RETRASO model was used to simulate the treatment performance and reactive transport in a tropical HSSF-CW.

A mechanistic approach, although physically sound and straightforward, requires a great deal of effort and skill to implement, as well as considerable computing resources for its application (Brooks and Tobias, 1996; Langergraber, 2011). A fundamental challenge for wetland scientists is determining the appropriate level of complexity for mechanistic models to effectively simulate the fate and behavior of target pollutants in wetlands (Min et al., 2011). Some investigators claim that simple models (closer to the empirical approach) have a higher practical value (Kadlec and Wallace, 2009), because mechanistic models are not able to explain the infinite complexity of the underlying phenomena, and that significant uncertainties are introduced in estimating a large number of model parameters. Others emphasize that more complex mechanistic models have a robust theoretical basis and are thus with better predictive potential (Min et al., 2011). Hence, the question is whether or not more complex, but also less manageable, models offer a significant advantage to the designer (Rousseau et al., 2004). To appreciate these aspects, the application of the CWM1-RETRASO model to simulate the reactive transport of a tropical pilot scale HSSF-CW system is compared to that with a semi-mechanistic biokinetic model implemented by Ojeda et al. (2008) in the RCB finite element code. The latter model features simplified descriptions of organic matter, nitrogen and sulphur transformation processes in HSSF-CWs, with fewer model parameters.

5.2   Methodology

5.2.1   2D simulation models

The two-dimensional simulation model CWM1-RETRASO is based on the work of Llorens et al. (2011a), in which the CWM1 model was implemented into the two-dimensional finite-element code RCB. Within the CWM1-RETRASO model, RCB provides the knowledge related to reactive transport and flow properties, while CWM1 provides the knowledge related to biochemical processes (Llorens et al., 2011a). In the model, physical oxygen transfer from the atmosphere to the water is included. Phosphorus transformations, biofilm growth, oxygen leaking from macrophytes and processes linked to clogging (i.e. solids accumulation) were not considered. In the RCB code, the concentration of primary and secondary species and aqueous complexes are expressed in mol/kg water (Saaltink et al., 2005). The molecular weight of O_2, is applied as the conversion factor from mg/l of COD and BOD to mol/kg units, whereas the molecular weight of N and S is used to convert the concentrations of nitrogen compounds and sulphate from mg/l to mol/kg units, respectively. The reaction rates are computed in per time (seconds) and expressed as mol/s kg water. For a detailed description of CWM1 implementation in RCB and utilization of these units, see Llorens et al. (2011a).

The CWM1-RETRASO and the semi- mechanistic model, developed in the work of Ojeda et al. (2008) have the same mathematical and functional form, but they compare and contrast in the following ways:

- CWM1-RETRASO uses Monod type expressions for substrate removal as a function of the bacterial concentrations. Six bacterial functional groups (facultative heterotrophs, autotrophs, fermenters, methanogens, sulfate reducers and sulfide oxidisers) are identified and associated with their respective organic matter transformation-degradation pathways in the CWM1-RETRASO model. In contrast, four simple generic microbial kinetic reaction equations are employed in the semi-mechanistic model to simulate microbial transformation and degradation processes for organic matter (aerobic respiration, denitrification, sulphate reduction and methanogenesis), without an explicit consideration of bacterial biomass growth. Bacterial biomass is regarded as not limiting the microbial kinetics. The organic matter degradation rates in the semi-mechanistic model are described by first-order kinetics with multiplicative saturation terms for the electron acceptors, and with multiplicative inhibition terms. This followed from the work of Van Cappellen and Gaillard (1996) that considered aerobic, anoxic and anaerobic degradation processes occurring at the same time in aquatic sediments (Van Cappellen and Gaillard, 1996).

- The explicit consideration of bacterial growth and lysis processes in the reaction scheme of the CWM1-RETRASO model makes it more detailed compared to the

86

reaction scheme of the semi-mechanistic model. Nevertheless, the RCB code architecture does not consider fixed biomass and therefore does not allow to conduct biofilm based-modeling. Hence, a constant bacterial concentration (estimated by fitting) is introduced into the influent and assumed to be transported with wastewater through the HSSF-CW bed similarly to the solute components and in this way maintaining an active microbial population within the filter.

There is a significant difference in the number of kinetic parameters, each of which has a specific mechanistic interpretation: CWM1-RETRASO needs 51, compared to 12 parameters for the semi-mechanistic model. The CWM1-RETRASO model is thus complex when focusing on the number of adjustable parameters, model variables and inputs.

- In the semi-mechanistic model, the reactions are assumed to use only dissolved organic matter as a substrate (measured in terms of dissolved COD), which is readily biodegradable by the four microbial reactions considered. An unspecific hydrolytic step to represent the conversion of influent particulate organic matter into dissolved organic matter is included. The CWM1-RETRASO model considers both dissolved and particulate organic matter. Furthermore, CWM1-RETRASO considers the non-biodegradable soluble and non-biodegradable particulate organic matter components in both the influent and effluent. This recognizes that part of the non-biodegradable organic matter in the system is debris associated with endogenous decay of bacterial biomass (Langergraber et al., 2009c; Llorens et al., 2011a).

- In both models, the slowly biodegradable fraction of COD (characterized as X_S and COD_X in CWM1-RETRASO and the semi-mechanistic model, respectively) represents the bulk of the biodegradable substrate and the most important input of hydrolysable substrate to the CW. Hydrolysis, which is an important process to initiate the degradation of slowly biodegradable substrate in wastewaters (Vymazal and Kröpfelová, 2009a), is formulated in CWM1-RETRASO by means of a surface-saturation-type of reaction, described in two reactions each one influenced by the heterotrophic bacteria (X_{HF}) or the fermenting bacteria (X_{FB}) (Llorens et al., 2011b). The hydrolysis by fermenting bacteria is assumed to be slower than by heterotrophic bacteria (Langergraber et al., 2009c). In the semi-mechanistic model, the hydrolysis process is described by first-order kinetics and a multiplicative exponential function that recognizes that most particulate organic solids are retained near the inlet.

5.2.2 Data for reactive transport simulation

The data used in this work were obtained from a pilot scale HSSF-CW located at the Jomo-Kenyatta University of Agriculture and Technology, Juja, Kenya (Mburu et al., 2013). Sampling was conducted between 2008 and 2011. The HSSF-CW consisted of three replicate wetland cells, each with an area of 22.5 m^2 (7.5 m x 3 m) and filled with granite type gravel (size 9 - 37 mm and porosity of 45 %) to a depth of 0.6 m. The primary effluent of a facultative pond treating domestic wastewater was introduced into the cells by continuous gravity feeding. The water depth was maintained at 0.5 m within the gravel bed with the aid of fixed outlet pipes. Data from two pilot cells, Cyp1 and Cyp2, planted with *Cyperus papyrus* was used. The data from the pilot cell Cyp2 were used for the model's calibration and those from the cell Cyp1 were used in the validation simulation. The mean hydraulic loading rates for the cell Cyp1 and cell Cyp2 during the study period were 195 ± 80 mm/d and 193 ± 86 mm/d, respectively. The mean theoretical hydraulic retention times were 1.6 ± 1.4 days in cell Cyp1 and 1.95 ± 1.0 days in cell Cyp2. The results obtained from the laboratory analysis of influent-effluent samples are summarized in Table 5.1.

Table 5.1. Influent and effluent characteristics in the period of the study (2008-2011)

| Quality Parameter | Unit | Influent | | | Effluent | | | |
		Range	Mean	n	Cyp1	n	Cyp2	n
DO	mgL^{-1}	0.21-18.5	6.32 ± 0.6	9	0.85 ± 0.12	5	0.72 ± 0.22	9
Temp	°C	19.4-26.6	23.3 ± 1.22	10	23.2 ± 1.7	5	22.6 ± 0.8	9
pH	pH	7.22-8.67	7.8 ± 0.2	10	7.04 ± 0.0	5	7.11 ± 0.11	9
NO$_3^-$-N	mgL^{-1}	0.1-3	1.1 ± 1.1	21	1.1 ± 0.8	16	0.9 ± 0.9	19
NH$_4^+$-N	mgL^{-1}	18-33	25.8 ± 4.5	19	18.8 ± 3.2	13	19.0 ± 5.8	19
BOD$_5$	mgL^{-1}	12-105	73.6 ± 17.7	10	34.6 ± 12.3	7	28.9 ± 9	9
COD	mgL^{-1}	52-354	159.5 ± 75.8	48	89.5 ± 45.1	22	91 ± 31.8	46
SO$_4^{2-}$	mgL^{-1}	18.5-100	66.7 ± 25.4	20	29.3 ± 12.2	17	20.1 ± 16.2	19

n: number of successfully analyzed samples

5.2.3 Hydraulic calibration

The definition of various aspects such as the spatial discretization of the wetland geometry (the grid), the initial water pressure in each node, the boundary conditions (inlet and outlet) and the hydraulic parameters (liquid longitudinal and transversal dispersion coefficients) is required to model the wetland hydraulics. The 2D grid consisted of 720 trapezoidal finite elements (60 columns and 12 rows) and 793 nodes. The dimensions of the grid were 7.5 m for the upper and lower bases (slopped at 1%), and 0.5 m and 0.575 m for the lateral sides at the inlet and the outlet, respectively. The partial pressures of the 793 nodes in the mesh were assigned after defining three outlet nodes, located at prescribed liquid pressures of 0.1064, 0.1058 and 0.1052 MPa yielding a mean water level of 0.489 m, which matched well with water depth observations at observation ports installed within the HSSF-CW bed.

The numerical hydraulic model was validated using a HRT distribution test with lithium in cell Cyp2. The tracer test conducted at an average flow rate of 2.54 m^3/d for 261 h yielded an experimental hydraulic retention time (HRT) of 70.2 h, whereas the simulated HRT was 71.6 h. The normalized variances of the experimental and simulated break through curves were 0.65 and 0.71, respectively. The difference in numerical and experimental tracer curves are attributed to the fact that RCB considers the HSSF-CW granular medium as homogeneous, ignoring filter medium heterogeneity and non-perfect subsurface flow conditions in the HSSF-CW. The liquid longitudinal and transversal dispersion coefficients were estimated by curve fitting in the RCB code, at 0.15 m and 0.07 m, respectively.

5.2.4 Reactive transport calibration, validation, inputs and model comparison

Biokinetic model calibrations and validations were performed against the observed concentrations of COD, NH_4^+-N, NO_3^--N and SO_4^{2-}-S in the effluent over five different influent wastewater flow rates and compositions (Table 5.2).

Table 5.2. Parameter set of the influent wastewater used for the calibration and validation simulations with the CWM1-RETRASO and semi-mechanistic model

Parameter (mg/l)	Flow rate (m³/d)				
	1.7 [2.3]	2.0 [2.2]	2.4 [2.7]	3.0 [4.0]	4.2 [5.3]
COD	127	256	96	157	141
BOD$_5$	85	64	66	105	80
NH$_4^+$-N	33	26	34	34	25.3
NO$_3^-$-N	0	0.2	0.1	0.1	2.1
SO$_4^{2-}$-S	37	68	93	68	100

[ ] Validation flow rate

To run CWM1-RETRASO, 19 inputs characterizing the influent (dissolved oxygen, COD fractions, NH$_4^+$-N, NO$_3^-$-N and SO$_4^{2-}$-S, initial bacterial concentrations, pH and alkalinity) and one input for the water flow rate and one for the hydraulic retention time are necessary. Fractionation of the influent wastewater COD was based on standard ratios given in the activated sludge models (Henze M. et al., 2000). The wastewater used in the pilot experiment was primary effluent with a wide range of the BOD$_5$:COD ratio (0.13 - 0.83) (Mburu et al., 2013) . The fractionation of COD was conducted as follows: S$_A$ (Acetate) = 15%, S$_I$ (Inert soluble COD) = 5%, X$_I$ (Inert particulate COD) =5%, X$_S$ (Slowly biodegradable particulate COD) = BOD$_{21}$-BOD$_5$ (BOD$_{21}$ approximated at 90% COD), and S$_F$ (Fermentable, readily biodegradable soluble COD) = BOD$_5$-S$_A$. Default values for the biokinetic model parameters as reported by Llorens et al. (2011b) were used in the simulations.

The calibration of the semi-mechanistic model involved the optimization (by trial and error as RCB does not provide tools for automatic parameter estimation using measured data) of the first order kinetic constants (found to be the most sensitive) for aerobic respiration, sulphate reduction and methanogenesis. Default model values for the half-saturation constants and inhibition constants were used. To run the model, 9 inputs characterizing the influent (dissolved oxygen, COD fractions, NH$_4^+$-N, NO$_3^-$-N and SO$_4^{2-}$-S, pH and alkalinity) and one input for water flow rate and one for the hydraulic retention time are required. The parameter and rate constants associated with the kinetic equations for organic matter degradation are available in Ojeda et al. (2008). As the reactions in the model are assumed to use dissolved COD as a substrate, the fractionation of the influent COD was implemented as: COD$_X$ (particulate slowly biodegradable COD) = 60% of COD, and CH$_2$O (dissolved biodegradable COD) = 40% of COD.

The results provided by the validation simulations were treated and used in the comparison of the organic matter transformation and degradation rates, as well as of the relative contribution to organic matter removal by the different microbial pathways in the model. To measure how well the models predict the observed effluent concentrations of the cell Cyp1 (validation simulations), the prediction residual sum of squares (PSS), defined as the sum of squared difference between the observed and predicted values (Cox et al., 2006), was used.

5.3   Results

5.3.1   Reactive transport simulations

Figure 5.1 shows the measured and simulated effluent concentrations of total COD in the calibration and validation simulations from CWM1-RETRASO (Top) and from the semi-mechanistic model (Bottom). The values were obtained with the defined initial bacterial concentrations for CWM1-RETRASO model (Table 5.3) and optimized kinetic constants for the semi-mechanistic model (Table 5.4). The simulated values obtained for the effluent COD corresponded well with the measured data. Accordingly, both models predicted the HSSF-CW performance well. However, the CWM1-RETRASO model showed a better predictive performance for COD (PSS=40.6) than the semi-mechanistic model (PSS=61.8).

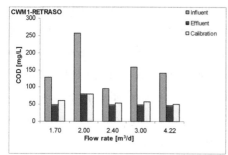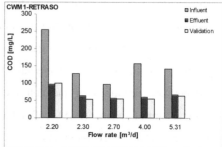

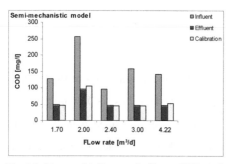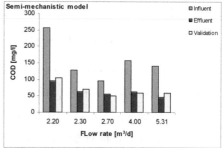

Fig. 5.1. Measured influent and effluent COD, and predicted effluent concentrations in the calibration and validation simulations from the CWM1-RETRASO (Top), and from the semi-mechanistic model (Bottom)

Table 5.3. Definition of initial bacterial concentrations for the CWM1-RETRASO model over five wastewater flow rates during calibration

Bacteria function group	Bacterial concentration (mg COD$_{BM}$/l)				
	1.7 m³/d	2.0m³/d	2.4m³/d	3.0m³/d	4.2 m³/d
Heterotrophic bacteria that consume S$_F$	44	113	111	100	100
Fermenting bacteria	435	226	283	339	305
Heterotrophic bacteria that consume S$_A$	44	113	111	100	100
Autotrophic nitrifying bacteria	10	10	10	10	40
Acetotrophic methanogenic bacteria	735	961	961	1130	960
Acetotrophic sulphate reducing bacteria	509	735	735	1074	735
Sulphide oxidising bacteria	2	2	2	2	5

S$_F$= Fermentable, readily biodegradable soluble COD
S$_A$= Fermentation products as acetate
BM=Bacterial biomass

Table 5.4. Values of the semi-mechanistic model kinetic parameters optimized in this work

First order kinetic rate constant (s^{-1})	Ojeda et al. (2008)	This work				
		1.7 m³/d	2.0m³/d	2.4m³/d	3.0m³/d	4.2 m³/d
Aerobic respiration ($k O_2$)	7.50E-06	7.50E-08	7.50E-06	7.50E-06	1.00E-05	1.00E-05
Sulphate reduction ($k SO_4^{2-}$)	3.00E-06	3.00E-06	3.00E-06	3.00E-06	3.60E-05	3.00E-05
Methanogenesis ($k CH_4$)	2.20E-06	2.20E-04	2.20E-06	2.20E-06	2.20E-05	2.20E-05

Figure 5.2a and 5.2b shows the measured influent and effluent, and simulated effluent concentrations of NH_4^+-N and SO_4^{2-}-S from the calibration and validation simulations using both models. The predicted effluent ammonia and sulphate concentrations fitted well with the experimentally determined effluent concentrations. The residual sum of squares from the validation simulations for the CWM1-RETRASO model were 21.0 and 77.3 for NH_4^+-N and SO_4^{2-}-S, respectively. Those for the semi-mechanistic model were 34.6 and 178.3, respectively. The CWM1-RETRASO showed a better predictive potential for NH_4^+-N and SO_4^{2-}-S, notwithstanding the fact that it entailed the fine tuning of different parameter sets to obtain a match of the effluent concentrations by both models.

A

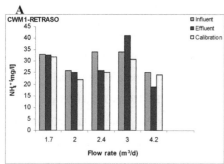

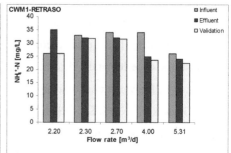

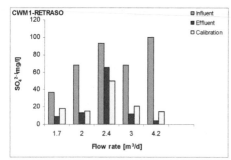

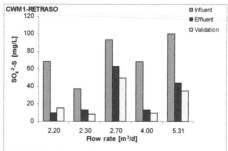

B

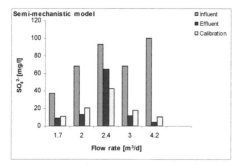

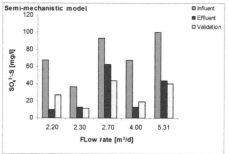

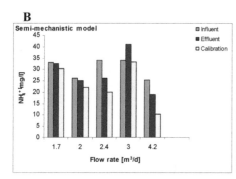

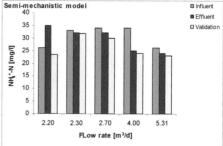

Fig. 5.2. (A and B): Influent, and measured and simulated NH_4^+-N and SO_4^{2-}-S effluent, concentrations from the calibration and validation with the CWM1-RETRASO (a) and the semi-mechanistic (b) model

5.3.2 Transformation and degradation reaction rates

For comparison of transformation and degradation rates in the kinetic pathways of the two biokinetic models, the validation results from simulations with the influent data at the flow rate of 2.2 m^3/d (Table 5.2) are presented in this section. The CWM1-RETRASO predicted effluent concentrations (in mg L^{-1}) were COD, 80.9; X_I, 34.3; S_I,12.1; X_S, 5.3; S_A, 20.7; S_F, 8.6; NH_4^+-N, 25.2; NO_3^--N, 0 and SO_4^{2-}-S, 15. The main part of the predicted effluent COD is inert, i.e. soluble inert (S_I) and particulate inert (X_I) COD. The semi-mechanistic model effluent concentrations (in mg L^{-1}) were COD, 104.7; COD_X, 46.5; CH_2O, 58.1; NH_4^+-N, 15.2; NO_3^--N, 0 and SO_4^{2-}-S, 27.1.

5.3.2.1 Hydrolysis and fermentation

The hydrolysis processes were found to take place mainly near the inlet zone of the simulated HSSF-CW (Fig. 5.3). The largest hydrolysis rates were found to be within the first meter of the wetland length. Maximum hydrolysis rates of 0.004 mol substrate/2.0E5 s·kg water and 0.001 mol substrate/2.0E5 s·kg water were observed in the simulation results for heterotrophic bacteria and fermenting bacteria, respectively, in the simulations with the CWM1-RETRASO model. A similar pattern of the hydrolysis activity (i.e. near the inlet) was observed with the semi-mechanistic model, achieving hydrolysis rates up to 0.006 mol substrate/2.0E5 s·kg water. A low residual concentration (< 20 mg/l) of the hydrolysable substrate (i.e. the slowly biodegradable COD fraction, X_S in the CWM1-RETRASO and COD_X in the semi-mechanistic) remained in the wetland towards the outlet. An internal production of readily soluble COD with a maximum concentration of 45 mg/l by the X_{HF} and 10 mg/l by the X_{FB} hydrolysis in CWM1-RETRASO and 81 mg/l in the semi-mechanistic model were simulated.

The fermentation process is observed to start away from the inlet and the top of the water column, but extending up to midway along the length of the simulated wetland (Fig. 5.3). The location of the highest activity of organic matter fermentation is observed to be near the output of the readily biodegradable COD (S_F) fraction by the hydrolysis process and near the inlet, where there is a direct input of S_F via the influent wastewater.

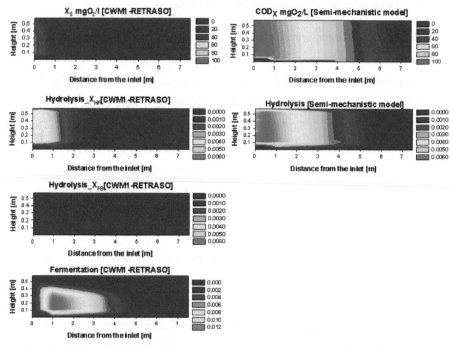

Fig. 5.3. Simulated changes along the wetland length of the concentration of slowly biodegradable substrate (X_S and COD_X) and of the rates [mol substrate/2.0E5 s·kg water] of the hydrolysis and fermentation processes with the CWM1-RETRASO (left) and semi-mechanistic (right) model at the wastewater flow rate of 2.2 m³/d

5.3.2.2 Aerobic and anoxic (denitrification) processes

Figure 4 shows the distribution patterns of the concentrations of dissolved oxygen (DO), ammonia and nitrates along the depth and length of the simulated HSSF-CW in the validation simulations using the CWM1-RETRASO and the semi-mechanistic models. The simulation of the oxygen distribution in the wetland shows that oxygen was limited to the inlet and the top of the water column, following the main sources of oxygen included in the CWM1-RETRASO and semi-mechanistic models namely, physical re-aeration and oxygen in the influent wastewater. The dissolved oxygen concentration was low in the influent and dropped rapidly within the first meter of the simulated HSSF-CW (Fig. 5.4). The ammonia concentration slightly decreased from the inlet to the outlet. The decrease especially coincided with regions of higher oxygen concentrations and potential for nitrification (at the surface of the water column in the wetland). Nitrates were simulated to be present at the influent of the

96

HSSF-CW and in regions with higher dissolved oxygen concentrations. However, the nitrate concentrations remained very low throughout the wetland (Fig. 5.4). Note that ammonia and nitrate plant uptake was not considered in the simulations.

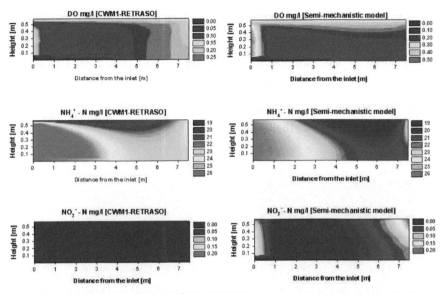

Fig. 5.4. Simulated changes of DO, NH_4^+-N and NO_3^--N concentrations along the length of wetland in the CWM1-RETRASO (left) and semi-mechanistic (right) model at the wastewater flow rate of 2.2 m^3/d

The simulation results showed that the aerobic processes took place close to the oxygen sources (Fig. 5.5). The highest nitrification rates were mainly simulated at the top of the water column (with oxygen transfer from the atmosphere) and the HSSF-CW inlet (with influent DO). The nitrification rates were very similar in both models in the order of magnitude 2E-5 and 6E-5 mol substrate/2.0E5 s·kg water (Fig. 5.5). However, the ammonium concentration did not vary significantly along the length of the wetland (Fig.4). Its removal was limited because the aerobic conditions necessary for nitrification were not spread enough throughout the wetland. In both models aerobic respiration (modeled as aerobic growth of heterotrophic bacteria X_{HA} and X_{HF} on acetate and on readily biodegradable organic matter respectively in the CWM1-RETRASO model) was only observed in a thin layer of water on the surface of the wetland and at the inlet (Fig. 5.5). In the simulations from both models, denitrification was located where nitrates were present, reaching a maximum especially near the inlet, due to the available influent nitrate concentration. Very low denitrification rates were detected along the rest of the length of the simulated wetland.

5.3.2.3 Anaerobic processes

The simulations of the anaerobic degradation of organic matter with both models showed that methanogenesis and sulphate reduction processes took place simultaneously at the same location within the wetland (Fig. 5.6). The sulphate reduction and methanogenesis rates showed the same pattern and the same order of magnitude in both models. The rates declined towards the outlet of the wetland which could be a result of substrate limitation.

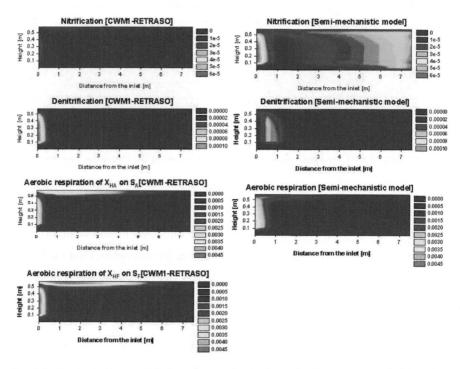

Fig. 5.5. Simulated changes of the process rates [mol substrate/2.0E5 s·kg water] of nitrification, denitrification and aerobic respiration along the wetland length in the CWM1-RETRASO (left) and semi-mechanistic (right) model at the wastewater flow rate of 2.2 m^3/d. X_{HA} refers to heterotrophic bacteria that consumes acetate (S_A) and X_{HF} refers to heterotrophic bacteria that consumes readily biodegradable organic matter (S_F) in the CWM1-RETRASO model

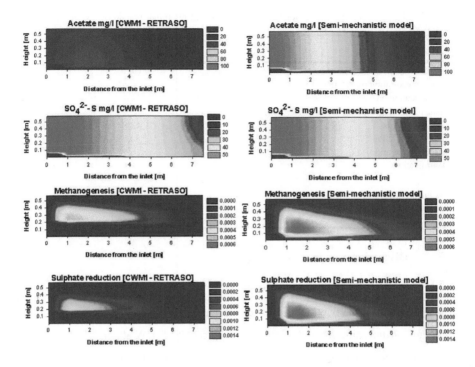

Fig. 5.6. Simulated changes of the concentration of acetate and sulphates and rates [mol substrate/2.0E5 s·kg water] of methanogenesis and sulphate reduction along the wetland length in the CWM1-RETRASO (left) and semi-mechanistic (right) model at the wastewater flow rate of 2.2 m³/d

5.4   Relative importance of the microbial reactions to organic matter degradation

The microbial pathways for organic matter degradation simulated in the two models contributed unevenly to organic matter removal in the simulated HSSF-CW (Table 5.5). Further, varying the wastewater flow rate at a constant influent concentration (as is the case in calibration-validation simulations) modified the relative contributions to COD removal by the microbial processes (Table 5.5), especially for the aerobic respiration, methanogenesis and sulphate reduction processes. Methanogenesis remained, however, the main organic matter removal pathway in the simulated HSSF-CW, contributing on average 37% and 49% COD removal in the simulations with the CWM1-RETRASO and semi-mechanistic model, respectively. In most of the simulations, the contribution of anaerobic bioconversion to organic matter removal was larger than that by anoxic and aerobic bioconversions.

Table 5.5. Relative average contribution (%) of the different microbial reactions to COD removal in the CWM1-RETRASO and semi-mechanistic model at the wastewater flow rate of 2.2 m³/d

	Model percentage [%]	
	CWM1	Semi-mechanistic Model
Aerobic respiration on S_F	19.3	N/A
Aerobic respiration on Acetate	19.1	N/A
Denitrification on S_F	0.3	N/A
Denitrification on Acetate	0.3	N/A
Methanogenesis	37.2	48.8
Sulphate reduction	23.9	34.6
Aerobic respiration	N/A	16.2
Denitification	N/A	0.3
Aerobic processes	38.4	16.2
Anoxic processes	0.6	0.3
Anaerobic processes	61	83.5

5.5 Discussion

5.5.1 Use of 2D mechanistic models for HSSF-CW simulation

The application of the two 2D mechanistic models to simulate transport and reactions of organic matter, nitrogen and sulphur in a tropical HSSF-CW attempts to overcomes the limitations of available empirical and first order models for constructed wetlands, i.e. modeling of contaminant degradation as a function of wetland type (subsurface or free water surface flow) and operating conditions (Bezbaruah and Zhang, 2009; Kadlec, 2000) by using rational models for both water flow and the biochemical reactions. In the CWM1-RETRASO model, the flow field is first calculated over a grid that describes in detail the geometry of the wetland and the hydrodynamic behavior of the wastewater flow through the porous media. The calculated flow variables are then used in computing the reactive transport in the wetland (Saaltink et al., 2005). The actual flow conditions in the wetland are approximated by using the residence time distribution obtained by experimental tracer test data. This approach considers the non-ideal flow conditions in HSSF-CW caused by effects of longitudinal dispersion, short-circuiting, dead zones and roots resistance in the reaction volume of the HSSF-CW bed (Toscano et al., 2009). However, the RCB code is not able to perfectly reproduce the receding limb of the experimental tracer curve, that should characterize the non-ideal flow in the porous HSSF-CW bed (Llorens et al., 2011a)

5.5.2 Reactive transport simulation and relative importance of microbial reaction pathways

5.5.2.1 Bacterial concentration

In both models, the wastewater reactive transport in the pilot HSSF-CW is governed by biodegradation. The CWM1-RETRASO model adjustment to experimental data (calibration) yielded bacterial concentrations (Table 5.3) that are the initial bacterial concentrations treated as a constant suspended biomass input with the influent wastewater. This is necessary because RCB's architecture does not consider fixed biomass. The consideration of growth and decay of biomass in suspension is acceptable as the biofilms growing in HSSF-CWs are typically very thin and diffusion limitations in the biofilm can be omitted (García et al., 2010). However, Langergraber and Šimůnek (2012) showed by means of simulations conducted in software with fixed biomass (i.e. biofilm-based modeling capability), together with root oxygen release from wetland plants that the bacteria concentration profiles in CWs are different from those simulated with a constant concentration of influent suspended bacterial biomass as is the case with the CWM1-RETRASO. This could possibly distort the mechanistic interpretation of transformation and degradation patterns within the HSSF-CW bed. Further, whereas anaerobic processes predominate in subsurface flow systems, aerobic processes may be found in the proximity of wetland plants (Stottmeister et al., 2003). In all likelihood, oxygen release and hence redox potential and the diversity of the rhizosphere microbial community vary according to the macrophyte plant species and environmental conditions (Faulwetter et al., 2009; Llorens et al., 2011a). Results are, however, not consistent in the literature and no mathematical relationships have been developed so far. Thus, the non-inclusion of the plant processes into the models may cause some disagreements between the model output and experimental measurements, as some processes (e.g. evapotranspiration and nutrient uptake) have been reported to be significant in constructed wetlands (Kadlec and Wallace, 2009).

Indeed, the fact that the bacterial concentration was the only variable that required fine-tuning to achieve reactive calibration for the CWM1-RETRASO model in this study suggests the importance of biofilm-based modeling for subsurface flow constructed wetlands. This indeed represents important phenomena influencing microbial reactions such as diffusion limitation as well as the stratification of metabolic processes or bacterial profiles across the wetland bed present (Langergraber and Šimůnek, 2012). Of importance too is the fact that at the moment there have been very few investigations of the distribution of the microbial biomass (Krasnits et al., 2009) and the actual concentrations of the specific bacterial groups that have been defined in biokinetic models, e.g., in CWM1 (Langergraber, 2011). Hence, a numerical rather than experimental method of evaluation of the kinetic parameters and variables for treatment processes should be acceptable.

The bacterial concentration showed a dominance of the anaerobic acetotrophic methanogenic bacteria, acetotrophic sulphate reducing bacteria and fermenting bacteria, which seemed to thrive in the oxygen deficient HSSF-CW. The heterotrophic bacteria (which are facultative) maintained a modest population, while the sulphide oxidising bacteria and the autotrophic nitrifying bacteria were calibrated to be present in very low quantities. They may have been outcompeted by heterotrophs for oxygen. The estimated values for the initial bacterial concentration are comparable to and within the range of those determined from simulation studies of subsurface constructed wetlands through a range of temperature, e.g. Llorens et al. (2011a, b), Mburu et al. (2012) and Rousseau (2005). Nevertheless, the simulations did not show a consistent bacterial concentration trend vis a vis the wastewater flow rates (Table 5.3), which seems to be due to the varying concentration of components in the influent wastewater arriving at the HSSF-CW from the facultative pond. Additionally, Krasnits et al. (2009) reported no significant differences between seasons in microbial community distribution in a HSSF-CW. Rather, depth was found to have a greater influence on the distribution of microbial communities. Hence, for the good performance of HSSF-CW, the design and operation of the wetland such as to afford enough contact time with microbial communities that proliferate in the biofilms on the surface of the porous media remains key, rather than the consideration of climate and seasonal effects.

5.5.2.2   Model predictions

The model predictions were found to be in quantitative agreement with the set of experimental data (Fig. 5.1 and 5.2). However, in all the validation simulations the CWM1-RETRASO model showed a better predictive performance compared to the semi-mechanistic model according to the prediction residual sum of squares. Furthermore, the CWM1-RETRASO model demonstrated in a better way the interaction of components, i.e. concentrations of microbial functional groups and their competition for substrates and the formation of intermediary products within the wetland (Fig. 5.5 and 5.6). Thus, the CWM1-RETRASO model meets in a better way the need to analyze and understand the biochemical environment in the HSSF-CW bed and its relation with the removal of organic matter, ammonia and sulphate. Also it has shown to be a useful tool in elucidating quantities (estimates of bacterial concentrations) which could not directly be identified from the observed concentration data. On the other hand, the semi-mechanistic model was not transparent (mechanistic) enough since the first order rate constant represents a lumped interaction of substrates and electron acceptors in a microbial catalyzed redox reaction.

The simulation profiles along the length and depth of the HSSF-CW demonstrated that the organic matter transformation and biodegradation reactions took place at the same time, in different locations (aerobic, anoxic and anaerobic), while others occurred in sequence (nitrification-denitrification) or in parallel (fermentation-methanogenesis-sulphate reduction) (Fig. 5.5 and 5.6). This agrees with the literature descriptions of simultaneous co-existence of areas with different redox status in the HSSF-CW bed (Faulwetter et al., 2009).

According to the simulations, the highest rates of organic matter transformation (hydrolysis and fermentation (Fig.5.3)) and removal (aerobic respiration, methanogenesis, nitrate and sulphate reduction (Fig. 5.5 and 5.6)) occurred in the first sections or near the inlet of the evaluated HSSF-CW. This can be associated with the availability of ample substrate at this section (directly fed through the influent); ultimately, the substrate concentration profiles and removal rates fall towards the outlet of the wetland (Trang et al., 2010). Simulations by both models showed the maximum aerobic respiration rates to be larger than those observed for anaerobic respiration processes, due to the large aerobic microorganisms growth rates (in CWM1-RETRASO) and first order aerobic respiration rates (in semi-mechanistic model) compared to the corresponding anaerobic values. For example the maximum rates of aerobic and anaerobic respiration (methanogenesis and sulphate reduction) on acetate are 0.004, 0.0005 and 0.0008 mol substrate/2.0E5 s·kg water, respectively, in the CWM1-RETRASO model (Fig. 5.5 and 5.6). Nevertheless, because they are widespread in the HSSF-CW bed, anaerobic biochemical reactions involved in organic matter degradation are the most important (Caselles-Osorio et al., 2007; García et al., 2007). The significant contribution of the methanogenic pathway in organic matter degradation (Table 5.5) in the simulated HSSF-CW has been observed in a similar mechanistic modeling work with the CWM1-RETRASO model (Llorens et al., 2011b) and through experimental work with mass balance and stoichiometric calculations (Caselles-Osorio et al., 2007)

The low DO concentrations simulated within the wetland and also observed in the effluent measurements from the pilot HSSF-CW indicate the presence of reducing conditions (Camacho et al., 2007). This circumstance explains the relatively small area occupied by aerobic respiration activity compared to that occupied by anaerobic processes in the simulated HSSF-CW. Anaerobic processes were located in most of the wetland in agreement with literature observations on HSSF-CWs in which aerobic processes only predominate near oxygen sources (roots and on the rhizoplane), whereas in the zones that are largely oxygen free, anaerobic processes such as denitrification, sulfate reduction and/or methanogenesis take place (Kadlec and Wallace, 2009; Vymazal, 2005). Indeed, HSSF-CW systems are generally considered to be anaerobic treatment systems as they have a limited potential for aerobic and anoxic conditions but rather strong reducing conditions often prevail (García et al., 2010; Vymazal, 2005). The specific mass balance for the removal pathways showed that the contribution to organic matter removal by anaerobic processes remained higher than that by anoxic and aerobic processes (Table 5.5). Aerobic respiration of organic matter and the biological oxidation of ammonium to nitrate (for anoxic respiration) with nitrite as an intermediate requires oxygen (García et al., 2010; Vymazal, 2007; Vymazal and Kröpfelová., 2009; Vymazal and

Kröpfelová, 2009b) which was the most limiting substrate in the simulated HSSF-CWs.

5.6 Biokinetic model comparison

The semi-mechanistic model provides an opportunity to inform by comparison how well the CWM1-RETRASO model supports the understanding and prediction of biochemical transformation and degradation processes in constructed wetlands, and the effort together with the computing resources needed in applying the fully mechanistic CWM1-RETRASO model. The modeling of the wetland hydraulics is the same in both models using a finite element scheme in the RCB software, which also provides a spatially and temporally resolved process specific mass balance for reactive species (Llorens et al., 2011b; Ojeda et al., 2008; Saaltink et al., 2005). Thus, the models can predict time and space dependent concentrations, a fact that makes the evaluation and comparison of both biokinetic models easy, by comparing the distribution of rates of the different reactions involved in organic matter transformation and degradation along the depth and length of the HSSF-CW bed. In both models the organic matter transformation and degradation is influenced by the availability of oxygen, nitrates and sulphates, reflecting the potential of aerobic, anoxic and anaerobic conditions in the HSSF-CW. The acceptable quantitative performance of the semi-mechanistic model is judged interesting because with a simplified but robust biokinetic model it is found sufficient to perform the task (effluent prediction) comparatively well. Generally, this may assist in the identification of potentially unnecessary model complexity in the CWM1-RETRASO model, which may be vital to allow for:

1) Including other important dynamic effects, such as evapotranspiration, sorption, plant processes, biofilm development and media clogging without complicating the model further. Non-inclusion of these processes in the CWM1-RETRASO model limits its application as a simulation tool for design.

2) Avoiding the short-comings with high complexity in models, which increases the number of unknown parameters and the possible dependencies between them. These factors make the accurate estimation of parameter values very difficult.

Without employing the rigorous parameters intensive microbial growth-lysis kinetics implemented within the CWM1-RETRASO model, the semi-mechanistic model was found computationally less demanding and more easy to run because of the fewer processes, components and parameters to be calibrated and computed. This can be seen from the simulation time difference between CWM1-RETRASO and the semi-mechanistic model: by a factor of up to 12, with a Duo CPU T9400 @ 2.53 GHZ and 3.48 GB RAM PC. This time factor can lead to simulation times deemed too long to permit practical use of the model for standard routine simulation purposes.

As it takes quite some time and experience to be able to produce realistic simulation results with CW mechanistic simulation models (Langergraber, 2011), it is imperative to differentiate modeling approaches for routine design simulation on the one hand and the detailed exploration simulation of CWs. Modeling to optimize design and operation requires a reasonable balance between a detailed description and practical handling of CW systems (Langergraber et al., 2009b). On the other hand, it is important that constructed wetland mechanistic models produce a proportional insight into the factors affecting pollution removal visa vis the difficulty of estimating a large number of parameters (Marsili-Libelli and Checchi, 2005).

5.7   Conclusion

The importance of mechanistic models in simulating and providing insights into the microbial processes involved in organic matter transformation and degradation in HSSF-CWs has been demonstrated with the CWM1-RETRASO model using HSSF-CW performance data from the tropics. Simulated effluent COD, NH_4^+-N and SO_4^{2-}-S concentrations showed a reasonably good fit to the measured concentrations across different pollutant loading rates and hydraulic retention times. The calibration of the bacterial populations demonstrated significant influence on the model's performance. The model correctly showed that anaerobic degradation is the dominant mechanism of organic matter removal in HSSF-CW. A comparison of the CWM1-RETRASO model with an alternative, less complex model was useful to evaluate the calibration effort and computational time required with a fully mechanistic model. The CWM1-RETRASO model was found suitable for simulation aimed for a better understanding of the CW system transformation and degradation processes. The model was, however, found expensive in computing and time resources required to perform the simulations.

5.8   References

Bezbaruah AN, Zhang TC. pH, redox, and oxygen microprofiles in rhizosphere of bulrush (Scirpus validus) in a constructed wetland treating municipal wastewater. Biotechnology and Bioengineering 2004; 88: 60-70.

Bezbaruah AN, Zhang TC. Incorporation of oxygen contribution by plant roots into classical dissolved oxygen deficit model for a subsurface flow treatment wetland. Water
Science and Technology 2009; 59: 1179-1184.

Brooks RJ, Tobias AM. Choosing the best model: Level of detail, complexity, and model performance. Mathematical and Computer Modelling 1996; 24: 1-14.

Camacho JV, Martinez ADL, Gomez RG, Sanz JM. A comparative study of five horizontal subsurface flow constructed wetlands using different plant species for domestic wastewater treatment. Environmental Technology 2007; 28: 1333-1343.

Caselles-Osorio A, Porta A, Porras M, García J. Effect of High Organic Loading Rates of Particulate and Dissolved Organic Matter on the Efficiency of Shallow Experimental Horizontal Subsurface-flow Constructed Wetlands. Water, Air, & Soil Pollution 2007; 183: 367-375.

Cox GM, Gibbons JM, Wood ATA, Craigon J, Ramsden SJ, Crout NMJ. Towards the systematic simplification of mechanistic models. Ecological Modelling 2006; 198: 240-246.

Faulwetter JL, Gagnon V, Sundberg C, Chazarenc F, Burr MD, Brisson J, et al. Microbial processes influencing performance of treatment wetlands: A review. Ecological Engineering 2009; 35: 987-1004.

García, Capel V, Castro A, Ruíz I, Soto M. Anaerobic biodegradation tests and gas emissions from subsurface flow constructed wetlands. Bioresource Technology 2007; 98: 3044-3052.

García J, Rousseau DPL, MoratÓ J, Lesage ELS, Matamoros V, Bayona JM. Contaminant Removal Processes in Subsurface-Flow Constructed Wetlands: A Review. Critical Reviews in Environmental Science and Technology 2010; 40: 561-661.

Hench KR, Bissonnette GK, Sexstone AJ, Coleman JG, Garbutt K, Skousen JG. Fate of physical, chemical, and microbial contaminants in domestic wastewater following treatment by small constructed wetlands. Water Research 2003; 37: 921-927.

Henze M., Gujer W., T. M, M. vL. Activated sludge models ASM1, ASM2, ASM2d and ASM3. Scientific and Technical Report No. 9. . IWA Publishing, London, UK. 2000.

Kadlec. The inadequacy of first-order treatment wetland models. Ecological Engineering 2000; 15: 105-119.
Kadlec, Wallace S. Treatment wetlands. 2nd ed. Boca Raton, Fla: CRC Press, 1048 pp. 2009.

Kivaisi AK. The potential for constructed wetlands for wastewater treatment and reuse in developing countries: a review. Ecological Engineering 2001; 16: 545-560.

Krasnits E, Friedler E, Sabbah I, Beliavski M, Tarre S, Green M. Spatial distribution of major microbial groups in a well established constructed wetland treating municipal wastewater. Ecological Engineering 2009; 35: 1085-1089.

Langergraber G. Simulation of the treatment performance of outdoor subsurface flow constructed wetlands in temperate climates. Science of the Total Environment 2007; 380: 210-219.

Langergraber G. Numerical modelling: a tool for better constructed wetland design? Water Science and Technology 2011; 64: 14-21.

Langergraber G, Giraldi D, Mena J, Meyer D, Pena M, Toscano A, et al. Recent developments in numerical modelling of subsurface flow constructed wetlands. Science of the Total Environment 2009a; 407: 3931-3943.

Langergraber G, Giraldi D, Mena J, Meyer D, Peña M, Toscano A, et al. Recent developments in numerical modelling of subsurface flow constructed wetlands. Science of the Total Environment 2009b; 407: 3931-3943.

Langergraber G, Rousseau DPL, Garcia J, Mena J. CWM1: a general model to describe biokinetic processes in subsurface flow constructed wetlands. Water Science and Technology 2009c; 59: 1687-1697.

Langergraber G, Šimůnek J. Reactive Transport Modeling of Subsurface Flow Constructed Wetlands Using the HYDRUS Wetland Module. Vadose Zone Journal 2012; 11.

Liolios KA, Moutsopoulos KN, Tsihrintzis VA. Modeling of flow and BOD fate in horizontal subsurface flow constructed wetlands. Chemical Engineering Journal 2012; 200–202: 681-693.

Llorens, Saaltink MW, García J. CWM1 implementation in RetrasoCodeBright: First results using horizontal subsurface flow constructed wetland data. Chemical Engineering Journal 2011a; 166: 224-232.

Llorens, Saaltink MW, Poch M, García J. Bacterial transformation and biodegradation processes simulation in horizontal subsurface flow constructed wetlands using CWM1-RETRASO. Bioresource Technology 2011b; 102: 928-936.

Marsili-Libelli S, Checchi N. Identification of dynamic models for horizontal subsurface constructed wetlands. Ecological Modelling 2005; 187: 201-218.

Mburu N, D. Sanchez-Ramos, Diederik P.L. Rousseau, J.J.A van Bruggen, George Thumbi, Otto R. Stein, et al. Simulation of carbon, nitrogen and sulphur conversion in batch-operated experimental wetland mesocosms. Ecological Engineering 2012.

Mburu N, Tebitendwa S, Rousseau D, van Bruggen J, Lens P. Performance Evaluation of Horizontal Subsurface Flow–Constructed Wetlands for the Treatment of Domestic Wastewater in the Tropics. Journal of Environmental Engineering 2013; 139.

Min J-H, Paudel R, Jawitz JW. Mechanistic Biogeochemical Model Applications for Everglades Restoration: A Review of Case Studies and Suggestions for Future Modeling Needs. Critical Reviews in Environmental Science and Technology 2011; 41: 489-516.

Moutsopoulos KN, Poultsidis VG, Papaspyros JNE, Tsihrintzis VA. Simulation of hydrodynamics and nitrogen transformation processes in HSF constructed wetlands and porous media using the advection-dispersion-reaction equation with linear sink-source terms. Ecological Engineering 2011; 37: 1407-1415.

Ojeda E, Caldentey J, Saaltink MW, Garcia J. Evaluation of relative importance of different microbial reactions on organic matter removal in horizontal subsurface-flow constructed wetlands using a 2D simulation model. Ecological Engineering 2008; 34: 65-75.

Reddy KR, D'Angelo EM. Biogeochemical indicators to evaluate pollutant removal efficiency in constructed wetlands. Water Science and Technology 1997; 35: 1-10.
Rousseau DPL. Performance of constructed treatment wetlands: model-based evaluation and impact of operation and maintenance. PhD Thesis, Ghent University, Ghent, Belgium (available from http://biomath.ugent.be/publications/download/). 2005.

Rousseau DPL, Vanrolleghem PA, De Pauw N. Model-based design of horizontal subsurface flow constructed treatment wetlands: a review. Water Research 2004; 38: 1484-1493.

Saaltink, Carlos Ayora, Olivella S. User's guide for RetrasoCodeBright (RCB). Department of Geotechnical Engineering and Geo-Sciences, Technical University of Catalonia (UPC), Barcelona, Spain 2005.

Stottmeister U, Wießner A, Kuschk P, Kappelmeyer U, Kästner M, Bederski O, et al. Effects of plants and microorganisms in constructed wetlands for wastewater treatment. Biotechnology Advances 2003; 22: 93-117.

Toscano A, Langergraber G, Consoli S, Cirelli GL. Modelling pollutant removal in a pilot-scale two-stage subsurface flow constructed wetlands. Ecological Engineering 2009; 35: 281-289.

Trang NTD, Konnerup D, Schierup H-H, Chiem NH, Tuan LA, Brix H. Kinetics of pollutant removal from domestic wastewater in a tropical horizontal subsurface flow constructed wetland system: Effects of hydraulic loading rate. Ecological Engineering 2010; 36: 527-535.

Van Cappellen P, Gaillard J-F. Biogeochemical dynamics in aquatic sediments. Reviews in Mineralogy and Geochemistry 1996; 34: 335-376.

Vymazal. Horizontal sub-surface flow and hybrid constructed wetlands systems for wastewater treatment. Ecological Engineering 2005; 25 478–490.

Vymazal. Removal of nutrients in various types of constructed wetlands. Science of the Total Environment 2007; 380: 48-65.

Vymazal, Kröpfelová L. Removal of organics in constructed wetlands with horizontal sub-surface flow: A review of the field experience. Science of the Total Environment 2009a; 407: 3911-3922.

Vymazal, Kröpfelová. Removal of nitrogen in constructed wetlands with horizontal sub-sureface flow: a review. Wetlands 2009; 29: 1114-1124.

Vymazal J, Kröpfelová L. Removal of Nitrogen in Constructed Wetlands with Horizontal Sub-Surface Flow: A Review. Wetlands 2009b; 29: 1114-1124.

Wu S, Jeschke C, Dong R, Paschke H, Kuschk P, Knöller K. Sulfur transformations in pilot-scale constructed wetland treating high sulfate-containing contaminated groundwater: A stable isotope assessment. Water Research 2011; 45: 6688-6698.

Chapter 6: Simulation of carbon, nitrogen and sulphur conversion in batch-operated experimental wetland mesocosms

This chapter has been published as: Mburu, N., Sanchez-Ramos, D., Rousseau, D.P.L., van Bruggen, J.J.A., Thumbi, G., Stein, O.R., Hook, P.B., Lens, P.N.L., 2012. Simulation of carbon, nitrogen and sulphur conversion in batch-operated experimental wetland mesocosms. Ecological Engineering 42(0) 304-315.

Abstract

A simulation model based on Constructed Wetland Model No. 1 (CWM1) using the AQUASIM mixed reactor compartment as a platform was built to study the dynamics of key processes governing COD and nutrient removal in wetland systems. Data from 16 subsurface-flow wetland mesocosms operated under controlled greenhouse conditions with three different plant species (*Typha latifolia, Carex rostrata, Schoenoplectus acutus*) and an unplanted control were used for calibration and validation in this mechanistic model. Mathematical equations for plant related processes (growth, physical degradation, decay, and oxygen leaching), physical re-aeration, as well as adsorption and desorption processes for COD and ammonium were included and implemented alongside CWM1 in the AQUASIM software, while some CWM1 parameters were adjusted to better fit the model predictions to experimental data during calibration. The simulation results showed that the model was able to describe the general trend of COD (R^2=0.97-0.99), ammonium (R^2=0.85-0.97) and sulphate (R^2=0.71-0.93) removal in the wetland mesocosms and also in their controls (unplanted) through the experimental temperature range of 12°C - 24°C. Oxygen transfer by physical re-aeration was found to be 0.05 and 0.09 g m^{-2} day^{-1} at 12 °C and 24°C respectively. The amount of root oxygen transfer was the highest for the planted mesocosms at 12°C at rates of 1.91, 0.94, and 0.45 g m^{-2} day^{-1} in the *Carex, Schoenoplectus* and *Typha* mesocosmmesocosms respectively, indicating that COD of the bulk wastewater was removed mainly by anaerobic processes under the specific experimental situations. Measured COD removal was better in the planted mesocosms than in the control; differences were effectively modeled by varying the bacteria concentration. The sorption process was found to be important in simulating COD and ammonia removal under these experimental conditions.

6.0   Introduction

Constructed wetlands (CWs) are an increasingly important technical option for wastewater treatment and re-use in both developing and developed countries (Haberl 1999; Kivaisi 2001; Solano et al. 2004). They are potential alternative sanitation systems as they are treatment cost-effective with a potential for resource re-use and recovery (Puigagut et al. 2007). However, the technical development of these systems in terms of design guidelines is limited as the design equations of these systems are still based on empirical rules of thumb and or simple first-order decay models (Kadlec 2000; Moutsopoulos et al. 2011; Rousseau et al. 2004). These design models, mainly based on input-output data from CWs, provide limited insights into the performance aspects of CWs. The models are significantly site specific as they have parameters that have been derived from experiments with pilot CWs. Thus, the parameters are only valid for the specific boundary conditions for which they have been obtained. This defeats the goal of sound designs and successful replication and optimization of CW systems, an important aspect if the systems are to have public and institutional support. Consequently, the desire for a model that can be applied widely to various conditions encountered in the design and evaluation of constructed wetland systems has recently led to the pursuit of process-based models that describe the main processes in constructed wetlands in detail.

By describing the pollutant transformation and elimination processes taking place within the CWs, coupled to the hydraulic behavior of the CW's flow field, process-based or mechanistic models are a promising tool for understanding the parallel processes and interactions occurring in wetlands (Kumar and Zhao 2011; Langergraber 2007, 2008; Llorens et al. 2011a). This is anticipated to promote an increased understanding of the performance aspects and a sound conceptualization and design of constructed wetland systems. One such processes based model is the Constructed Wetland Model No.1 (CWM1) whose formulation is based on previous experiences of modeling processes in subsurface flow constructed wetlands as reviewed in Langergraber (2008). CWM1 is a general biokinetic model to describe biochemical transformation and degradation processes for organic matter, nitrogen and sulphur in subsurface flow constructed wetlands. The mathematical structure of CWM1 is based on the mathematical formulation of Activated Sludge Models (ASMs) as introduced by the IWA task group on mathematical modeling for design and operation of biological wastewater treatment. The reaction scheme, rate equations and kinetic constants of CWM1 describe aerobic, anoxic and anaerobic processes, and is therefore applicable to both horizontal and vertical flow constructed wetland systems; however other extra processes including porous media hydrodynamics (effect of dispersion, heterogeneity and dead zones), the influence of plants, the transport of particles/suspended matter to describe clogging processes, adsorption and desorption processes and physical re-aeration must be considered for the formulation of a full model of constructed wetlands (Langergraber et al. 2009; Llorens et al. 2011b).

The main scope of this study was to implement CWM1 in the software for identification and simulation for aquatic systems, AQUASIM (Reichert 1995), to simulate carbon, nitrogen and sulphur removal in constructed wetland systems. Besides the biochemical transformation and degradation processes described in CWM1, the influence of plants, physical re-aeration, adsorption and desorption processes have been considered. The model is used here to analyze the interactions between water, the granular substratum, macrophytes, and microorganisms for pollutant transformation and degradation, in batch-operated constructed wetland mesocosms, under a range of temperatures. Batch operation has the advantage of simplifying the hydraulics when integrating transport and transformation processes in porous media that are otherwise solved with advection-dispersion-reaction equations in 1D or 2D. The calibration of the CWM1-AQUASIM model is achieved here by adaptation of the model to fit actual data from constructed wetland mesocosm experiments, with the aid of sensitivity analysis and parameter estimation tools in AQUASIM. Biofilm development is not considered in the present work.

6.1 Materials and methods

6.1.1 The experimental constructed wetlands

The experimental constructed wetlands were operated under controlled greenhouse conditions at Montana State University in Bozeman, Montana, USA. Details of column design, construction and planting, as well as sampling and measurement, are fully described in Allen et al. (2002) and Stein et al. (2006). Briefly, 16 subsurface constructed wetland mesocosms were constructed from polyvinyl chloride (PVC) pipe (60 cm in height × 20 cm in diameter) and filled to a depth of 50 cm with washed pea-gravel (0.3–1.3 cm in diameter). Four columns each were planted with *Carex utriculata* (Northwest Territory sedge), *Schoenoplectus acutus* (hardstem bulrush) and *Typha latifolia* (broadleaf cattail), while four were left unplanted as controls. A series of 20-d incubations with artificial wastewater was conducted over 20 months at temperatures ranging from 4 to 24°C at 4°C steps. A synthetic wastewater simulating secondary domestic effluent was used with mean influent concentrations of 470 mg/l COD, 44 mg/l N (27 Org-N, 17 NH_4^+–N), 8 mg/l PO_4^{3-}–P, and 14 mg/l SO_4^{-2}–S. Columns were gravity drained 3 days prior to each incubation and then again at the start of each incubation. Upon each emptying, columns were refilled from above with new wastewater. Sampling from all 16 columns occurred at days 0, 1, 3, 6, 9, 14 and 20 of each incubation and those sub-samples were analyzed afterwards for the constituents.

6.1.2 Model description and implementation

Testing of the experimental data was performed by means of sensitivity analysis, parameter estimation and uncertainty analysis using AQUASIM computer program for the identification and simulation of aquatic systems developed by the Swiss Federal Institute for Environmental Science and Technology (Reichert 1998). The academic software AQUASIM is extremely flexible in allowing the user to specify transformation processes and to perform simulations for a user-specified model. The mixed reactor compartment configuration in AQUASIM was used and defined by the volume of the wetland mesocosm, active variables, active processes, initial conditions and inputs. Parameters in AQUASIM were set as fixed, dynamic state, list or formula variables. Dynamic processes were used for the growth and decay rate of all bacteria groups as included in the CWM1. When implementing CWM1 in AQUASIM, the subsurface-flow wetland mesocosms are assumed to act as single continuous stirred-tank reactor, supposing that all incoming constituents are evenly mixed throughout the entire mesocosm volume. This was corroborated by the fact that solution samples collected from 5, 15, and 30 cm depths during preliminary incubations showed no measurable vertical gradients for COD, dissolved organic carbon (DOC), or SO_4^{-2}–S (Allen et al. 2002). Default parameter values of CWM1 were adopted as published in Langergraber et al. (2009). Five plant processes (growth, physical degradation, decay, and oxygen leaching), physical re-aeration, as well as adsorption and desorption processes for COD and ammonium nitrogen, were also included as dynamic processes and are shown in Tables 1 and 2 in the appendix.

Plant growth is modeled by means of 'relative growth rates' as there are many data available in the literature. Plant growth is not zero-order, but depends on ammonium and nitrate concentrations (Rousseau 2005; Sanchez et al. 2004). Plant decay/senescence and physical degradation equations are based on the work of Wynn and Liehr (2001) and most importantly, plant material is no longer expressed as carbon but as COD, which is rather unusual but allows for a smooth integration with the COD-based microbial processes (Rousseau 2005). Adsorption of COD is modeled by a chemical non-equilibrium adsorption of the slowly biodegradable particulate COD, with a linear adsorption isotherm following the work of Henrichs et al. (2007). The time dependency of adsorption is described by the concept of two-site sorption.

116

Sorption is instantaneous on one part of the exchange sites whereas it is considered to be time-dependent on the remaining sites (Henrichs et al. 2007). Ammonium nitrogen sorption is described using a reversible Freundlich isotherm (McBride and Tanner 1999). Temperature impact on microbiological process rates and on plant growth is modeled via the Arrhenius relationship.

To run the model, 19 inputs characterizing the influent (Oxygen, COD fractions, N compounds, S compounds, and bacterial concentrations) and one input for water temperature are necessary. Fractionation of the influent wastewater COD was based on standard ratios given in the ASM models. Because the synthetic wastewater used in the experiment was mixed from sucrose, low molecular weight hydrolyzed meat protein and other chemicals for nutrients, the section for particulate inert organic particles was set to zero, therefore the fractionation was modified as follows: S_A (Acetate) = 10%, S_I (Inert soluble COD) = 4%, X_S (Slowly biodegradable particulate COD) = 61% of the measured influent COD, S_F (Fermentable, readily biodegradable soluble COD) = 25%. X_I (Inert particulate COD) was set to zero. S_O (dissolved oxygen) in the influent was estimated to be the saturation concentration at the set temperature. S_NO (Nitrate and Nitrite Nitrogen), and S_H$_2$S (Dihydrogensulphide sulphur) were assumed not to be present in the influent.

6.1.3   Sensitivity analysis

A sensitivity analysis was carried out to recognize the most important parameters influencing the prediction of carbon, nitrogen and sulphur concentrations and the growth of different microbial biomass in CWM1. In AQUASIM, the sensitivity analysis feature enables calculation of linear sensitivity functions of arbitrary variables with respect to each of the parameters included in the analysis (Reichert 1995). The sensitivity analysis results described in this study are those of the absolute-relative sensitivity function of AQUASIM (Eq. 6.1) that computes the absolute change in a model output variable, y, for a 100% change in any parameter of interest, p. This makes quantitative comparisons of the different parameters on a common variable possible.

$$\delta_{y,p}^{a,r} = p\frac{\partial y}{\partial p} \tag{6.1}$$

The uncertainty is determined by using the error propagation formula (Eq. 6.2), which is based on the linearized propagation of standard deviations of the parameters of interest, neglecting their correlation.

$$\sigma_y = \sqrt{\sum_{i=1}^{m}\left(\frac{\partial y}{\partial p}\right)^2 \sigma^2_{p_i}} \qquad (6.2)$$

Where p_i are the uncertain model parameters, σ_{p_i} are their standard deviations, $y(p_i.........p_m)$ is the solution of the model equations for a given variable at a given location and time, and σ_y is the approximate standard deviation of the model result. Identifiability of the model parameters was evaluated by use of parameter correlation matrix in AQUASIM.

6.1.4 Model calibration

Model parameters for CWM1 were adopted as described by Langergraber et al. (2009). The model requires the values of 51 kinetic parameters (16 first-order kinetic constants, 22 half-saturation coefficients and 13 inhibition constants) and 14 stoichiometric parameters (Llorens et al. 2011a). To optimize the parameter sets, the result of the sensitivity analysis was used to guide the selection and calibration of some kinetic coefficients. Data based on bulk measurements of COD, NH_4^+–N and SO_4^{-2}–S at 12°C, 16°C, 20°C and 24°C from the unplanted (control) mesocosms were used for calibration of the microbial pathways. At this time the matured experimental units had been receiving synthetic wastewater for 17 to 20 months. To adjust parameter values, the parameter estimation tool of AQUASIM (secant and simplex algorithms) and trial-and-error approach was used. The measured data of incubation at 12°C was used to estimate the initial bacterial concentration, the oxygen re-aeration coefficient, initial amount of sorbed ammonia and the COD adsorption parameters by simultaneous fitting of predicted COD, NH_4^+–N and SO_4^{-2}–S profiles to experimental data. Then, the final concentrations for each of the bacterial group at the end of 20 days were inputted as initial concentrations for the following 3-day refreshing incubation. Finally, the concentrations at the end of the 3-day refreshing incubation were used as the input for the next batch, i.e. incubation temperature.

6.1.5 Simulation and validation

The simulations were conducted with data from planted mesocosm, at 12°C, 16°C, 20°C and 24°C. Only the temperature, root oxygen loss parameter, ammonia and COD sorption parameters, and the initial bacterial concentration were changed in each simulation run. When estimating the initial bacterial concentrations, the lower limit values provided to AQUASIM were those obtained during calibration at the respective incubation temperatures with the unplanted mesocosms. For validation purposes, the model was run with a data set from a separate experimental campaign with higher sulphate concentrations in the mesocosms planted with *Schoenoplectus* (Stein et al. 2007).

6.2 Results

6.2.1 Sensitivity

The biokinetic parameters (kinetic rate constants, stoichiometric and composition parameters) and initial conditions were directly selected for sensitivity analysis. A sensitivity ranking of the most sensitive parameters according to the average absolute value of the "absolute relative" sensitivity function in AQUASIM is shown in Table 6.1. According to Table 6.1, the yield coefficients for methanogenic, sulphate reducing, fermenting and heterotrophic bacteria, saturation/inhibition coefficients for oxygen, sulphate, acetate, fermentable COD and hydrolysis, plant growth rate constant, rate constant for lysis of heterotrophic bacteria and the COD sorption coefficients are among the most sensitive and distinct parameters affecting the predicted concentrations.

Table 6.1. Sensitivity ranking and mean values of the relative sensitivity function of predicted effluent concentrations

Relative Sensitivity	S_A	S_F	S_NH	S_SO4	X_S
<1		K_SFB	m, Y_H, Kpl, μ_H		μ_H20, K_H, Y_H, b_H, K_X
>1-10<		Y_H, w, μ_H, k_S_COD, K_H, μ_FB, Y_FB,		Y_ASRB, μ_ASRB, μ_AMB, Y_AMB, K_SOASRB	
10>	μ_AMB, Y_AMB, μ_H, K_SAMB, K_OAMB				w, k_S_COD

S_A: Fermentation products as Acetate; S_F: Fermentable readily biodegradable soluble COD; S_NH: Ammonium and ammonia nitrogen; S_SO4: Sulphate sulphur; X_S: Slowly biodegradable particulate COD

6.2.2   Model calibration, identifiability and uncertainty

Default parameter values were implemented as proposed for CWM1 in Langergraber et al. (2009). Only 5 parameters needed to be adjusted during calibration as shown in Table 6.2. The correlation matrix (Table 6.2) shows minimal linear dependency between calibrated parameters, which increases their identifiability within the range of available data. It was found necessary to adopt low value for hydrolysis rate constant (through parameter estimation in AQUASIM) to be able to satisfactorily match the measured COD and NH_4^+ data. The remarkable difference between the hydrolysis rate constant value K_H in CWM1 and this work suggest low hydrolysis activity in the wetland mesocosm, which is conceivable considering that the constituents of the synthetic wastewater were mainly of low molecular weight. The estimated bacterial concentrations obtained during calibration at the different incubation temperatures, together with the percentage upper and lower bounds around the estimated bacteria concentrations are given in Table 6.3. It was found that acetotrophic methanogenic bacteria (X_AMB), fermenting bacteria (X_FB) and heterotrophic bacteria (X_H) become the most abundant organisms in the control mesocosm. The percentage concentration bounds suggest a low to medium uncertainty for the estimated initial bacterial concentrations. The correlation matrix (not shown) indicated a minimal linear correlation among the estimated initial bacteria concentration values, except for the heterotrophic and the acetotrophic methanogenic bacteria that had a significant negative correlation of 0.84. The goodness of fit between the observed and simulated concentrations in the calibrated model (control mesocosms) was evaluated with the coefficient of determination (R^2). The R^2 values for COD (0.97-0.99), NH_4^+–N (0.85-0.97) and SO_4^{-2}-S (0.71-0.93) indicate that the simulated data fit well with the observed data after calibration.

Table 6.2. Optimized parameters values and their correlation matrix

Parameters	Value in CWM1 (at 20°C)	Value in this work (at 20°C)	K_H	μ_ASRB	μ_FB	Y_AMB	Y_ASRB
K_H	3.00	0.58	1.00				
μ_ASRB	0.18	0.31	-0.18	1.00			
μ_FB	3.00	3.77	0.00	-0.33	1.00		
Y_AMB	0.03	0.04	-0.10	-0.13	-0.04	1.00	
Y_ASRB	0.05	0.04	-0.20	0.21	-0.47	-0.14	1.00

Table 6.3. The initial bacterial concentrations obtained in the model calibration with the upper and lower bounds around the estimated bacteria concentration (mgCOD/l)

Temp (°C)	X_Aini	X_AMBini	X_ASRBini	X_FBini	X_Hini	X_SOBini	Total conc.
12	67.1	58.8	14.4	1.2	363.3	2.9	507.6
16	79.4	58.3	6.7	55.4	138	2.1	339.8
20	37.9	71.3	7	37	83.7	2.7	239.6
24	35	95.3	5	50.3	45	1	231.6
% Upper bound	8.6	33.3	3.3	18.2	10	21.2	
% Lower bound	8.6	39.9	3.3	18.6	10	21.2	

6.2.3 Simulations results

Root oxygen transfer rates obtained from the model simulation are listed in Table 6.4. Oxygen transfer through diffusion or physical re-aeration were found to be at a low rate, calibrated at 0.05 g m^{-2} day^{-1} at 12°C and 0.09 g m^{-2} day^{-1} at 24°C in the control mesocosm. According to the simulations, appreciable plant root oxygen release occurred at 12°C at rates of 1.91, 0.94, and 0.45 g m^{-2} day^{-1} in the *Carex*, *Schoenoplectus* and *Typha* mesocosms, respectively. This can be explained by the fact that availability of oxygen is temperature dependent, since solubility of oxygen increases with decreasing temperature, which influences concentration gradients and internal transport of oxygen in the plants by molecular diffusion. Indeed besides the mesocosm planted with *Carex* at 16°C, no meaningful root oxygen leaching was obtained for all the tests in the planted columns at 16°C, 20°C and 24°C. The rate of root oxygen release decreased with an increase in temperature.

Table 6.4. Root oxygen transfer rates in the planted wetland mesocosms

Root oxygen transfer (as g O_2 per m^2 wetland area day^{-1})

	12^{0}C	16^{0}C	20^{0}C	24^{0}C
Carex	1.91	0.47	2.13E-05	2.65E-05
Schoenoplectus	0.94	2.60E-04	2.13E-05	2.66E-05
Typha	0.45	3.68E-02	5.38E-03	2.60E-04

Model simulations of COD, SO_4^{-2}-S and NH_4^+–N were compared to the observed data to demonstrate the degree of agreement and to discuss the most important phenomena visible in the data. Figures 6.2, 6.3 and 6.4 show the simulation profiles of COD, NH_4^+–N and SO_4^{-2}–S in the unplanted and planted wetland mesocosms at the various incubation temperatures. The simulation results demonstrated that the model was able to describe the general trend of COD, ammonium and sulphate removal in both planted and unplanted (control) mesocosms throughout the experimental temperature range of 12°C - 24°C. The estimated bacterial concentrations obtained during the simulations at the selected incubation temperatures, are given in Table 6.5. It was found that acetotrophic methanogenic bacteria (X_AMB), fermenting bacteria (X_FB) and heterotrophic bacteria (X_H) become the most abundant organisms in the simulated wetlands (planted mesocosms).

Table 6.5. Initial bacterial concentrations (mgCOD/l) values of planted wetland mesocosms defined during simulations

Set-up	Temp ($^\circ$C)	X_Aini	X_AMBini	X_ASRBini	X_FBini	X_Hini	X_SOBini	Total conc. (mgCOD/l)
				Bacteria group				
Carex	12	52.6	202.0	15.3	15.2	296.6	30.4	612.2
	16	79.5	185.8	6.7	56.5	138.0	17.0	483.5
	20	38.4	135.0	6.7	63.4	45.3	4.4	293.1
	24	35.0	95.3	5.0	50.3	45.0	1.0	231.6
Schoenoplectus	12	53.4	167.1	21.0	11.3	297.7	38.4	588.9
	16	79.4	131.1	6.7	55.4	138.0	9.0	419.6
	20	38.4	94.4	6.7	55.4	48.3	16.4	259.6
	24	10.0	110.1	6.7	55.4	68.2	10.8	261.2
Typha	12	52.8	102.6	16.7	10.2	297.1	82.3	561.7
	16	79.4	90.7	7.4	55.4	155.8	17.4	406.1
	20	42.0	103.0	6.7	43.8	86.4	2.4	284.2
	24	10.3	61.4	5.1	50.6	45.7	2.9	267.0

6.2.3.1 COD

Analyses of COD degradation pathways are shown in Fig. 6.1. The rates predicted by the model show that anaerobic degradation plays a major role on the COD removal both in the planted and unplanted mesocosm. Removal of COD through the aerobic pathway varied between 14.2 - 43.5 %, only 0 - 0.2 % was removed by anoxic reactions (denitrification), while 45.6 - 80.4 % of the COD removal was via the anaerobic pathway. The simulated residual was found to lie between 4.9 - 26.0 %. Larger removal of COD by aerobic processes (>30%) occurred at 12°C and 16 °C in the *Carex* and *Schoenoplectus* planted wetlands coinciding with higher root oxygen transfer rates as deduced in the simulation.

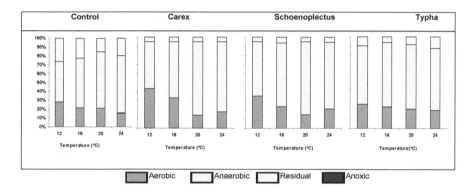

Fig. 6.1. Percentage contribution to COD degradation by aerobic, anoxic and anaerobic pathways from model simulations.

Fig. 6.2 compares simulated (with and without adsorption) and measured COD concentrations for both planted and unplanted mesocosms. The graphs show that COD predictions with adsorption on the gravel media were generally in good agreement with the measured data. However, in some instances there are moderate over/under estimations which seem to indicate that the mathematical model has a much faster/slower COD removal rate than experimental units. The difference between measured COD removal and simulated COD removal excluding adsorption is generally much larger than when adsorption is considered. This difference decreases with increase in the incubation temperature in the planted columns, however this trend is not observed in the simulated results of the unplanted columns. The adsorption pattern tends to follow that of the simulated bacterial concentrations. In practice it is often assumed that COD is adsorbed onto the biofilm covering the substrate from where it is further processed. Table 6.6 shows the coefficient of determination values (R^2) for the goodness of fit with and without adsorption processes during the simulation.

Table 6.6. R^2 for the goodness of fit for COD with (active) and without (inactive) COD adsorption processes

Temp(°C)	Control Active	Control Inactive	Carex Active	Carex Inactive	Schoenoplectus Active	Schoenoplectus Inactive	Typha Active	Typha Inactive
12	0.984	0.781	0.997	0.671	0.993	0.645	0.932	0.843
16	0.995	0.813	0.994	0.713	0.979	0.843	0.893	0.892
20	0.996	0.769	0.997	0.869	0.976	0.857	0.907	0.898
24	0.979	0.729	0.975	0.975	0.961	0.882	0.893	0.905

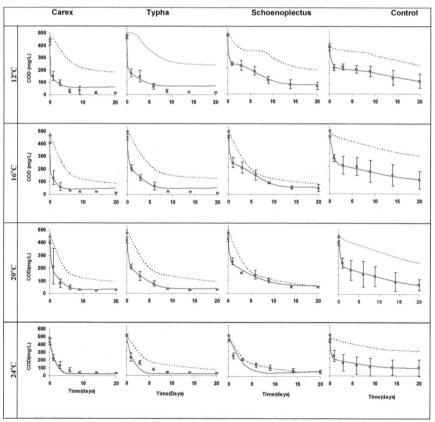

Fig.6.2. Simulated COD concentration compared with measured concentrations. Symbols are means (±S.D.) of observed concentrations from four replicates for each treatment. Solid lines are CWM1-AQUASIM simulation curves, dotted lines are CWM1-AQUASIM simulations without sorption.

124

6.2.3.2 SO₄⁻²–S

Measured and simulated SO_4^{-2}–S concentrations are compared in Fig.6.3. Model SO_4^{-2}–S predictions generally agree well with the measured data, except for some overestimation at the earlier days of incubation in some cases. The observed and simulated SO_4^{-2}–S removal profiles have got a pattern similar to those of COD i.e. a high removal rate within the first 3 days suggesting that sulphate was used as an electron acceptor for organic carbon removal.

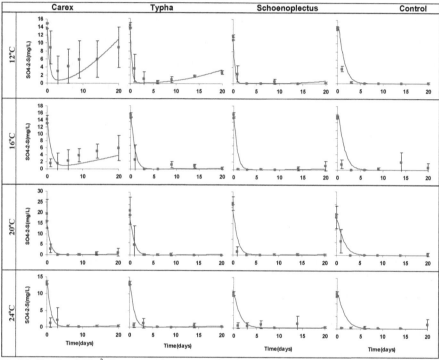

Fig.6.3. Simulated SO_4^{-2}–S concentration compared with measured concentrations. Symbols are means (±S.D.) of observed concentrations from four replicates for each treatment. Solid lines are CWM1-AQUASIM simulation curves.

6.2.3.3 NH₄⁺–N

Fig.6.4 shows the simulated NH_4^+–N concentration compared with the measured NH_4^+–N concentration in the wetland mesocosms. The observed rapid initial decline of NH_4^+–N in planted columns and increase of NH_4^+–N in unplanted columns are predicted well, and those can be achieved only with a sorption process included in the model. To do so, the process of reversible sorption of ammonium onto the gravel was

125

set to be active and an initial concentration of sorbed ammonium was also fitted. According to the simulations, the initial sorbed ammonium (X_NHini) concentration was found to be almost the same at different incubations for a specific column as shown in Table 6.7. This suggests that the full sorption potential of the gravel would be available after every 3 day refresh incubation, implying the ammonia adsorbed on the gravel was partly removed during the short period of draining and refilling of the wastewater between batches when the gravel was exposed to air. The variability of X_NHini between the mesocosms is explicable only as a combination of the root oxygen release potential of individual plant species and the below ground biomass characteristics i.e. root density, depth and thickness. The contribution by nitrification, plant uptake and the sorption processes to the observed decline in NH_4^+–N concentration in the planted mesocosms was computed as listed in Table 6.8.

Table 6.7. Initial concentration of adsorbed ammonia (g/Kg gravel)

Temp °C	Control	Carex	Schoenoplectus	Typha
12	13.9	3.2	5.0	6.0
16	14.0	3.0	5.4	5.8
20	14.3	3.0	6.0	6.0
24	13.8	3.0	6.7	6.0

Table 6.8. Percentage contribution to the observed decline in NH_4^+–N concentration by nitrification, plant uptake and sorption processes

Set-up	Temp°C	Nitrification	Plant Uptake	Sorption
Carex	12	2.0	20.5	70.2
	16	1.0	4.3	89.1
	20	0.1	4.0	89.8
	24	0.0	4.0	92.0
Schoenoplectus	12	0.1	30.0	55.0
	16	0.7	0.5	81.0
	20	0.1	0.3	74.7
	24	0.0	0.1	82.0
Typha	12	0.0	10.4	69.6
	16	0.0	1.5	75.0
	20	0.0	0.4	77.0
	24	0.0	0.3	75.0

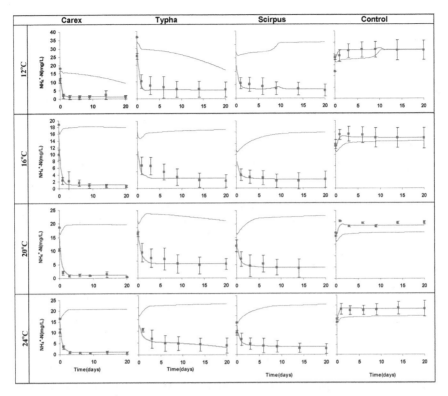

Fig. 6.4. Simulated NH_4^+–N concentration compared with measured concentration. Symbols are means (±S.D.) of observed concentrations from four replicates for each treatment. Solid lines are CWM1-AQUASIM simulation curves, dotted lines are CWM1-AQUASIM simulations without sorption.

6.2.3.4 Validation

For validation purposes, the model was run with a data set from a separate experimental campaign utilizing mesocosms planted with *Schoenoplectus* and higher sulphate concentrations (69.01 and 74.79 mg/l SO_4^{-2}–S) mesocosm (Stein et al. 2007). The simulated data fits reasonably well (Fig. 6.5) with R^2 values of (0.97, 0.92) and (0.64, 0.83) for COD and SO_4^{-2}-S at 14°C and 24°C respectively.

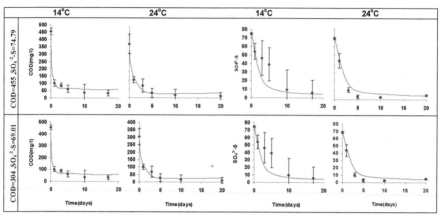

Fig.6. 5. Simulated COD and SO_4^{-2}–S concentration compared with observed concentration during model validation. Symbols are means (±S.D.) of observed concentrations from four replicates for each treatment. Solid lines are CWM1-AQUASIM simulation curves.

6.3   Discussion

Mechanistic models have become promising tools for understanding the parallel processes and interactions between water, granular media, macrophytes, litter, detritus and microorganisms occurring in constructed wetlands. The goal of this work was to implement within AQUASIM software the reaction model CWM1 and to yield simulation results comparable to concentrations actually measured for a constructed wetland mesocosm system operated in batch mode. The implementation of CWM1 with adsorption, physical oxygen re-aeration and plant processes into AQUASIM was done freely. This requirement of freedom in specifying a model in simulation software is essential to eliminate barriers for potential model users, while producing realistic simulation results (Meysman et al. 2003). Considering the large number of parameters in CWM1 defining different microbial pathways, colinearity prevented a high identifiability for some parameters of the six microbial groups considered in the model. Deviations between predicted and experimental data especially at low temperatures, may be caused by the non-identification of some CWM1 bacteria groups due to a compensation effect, underlining the need for higher resolution experimental data, for example a narrow difference between incubation temperatures at lower temperatures and/ or sampling time steps. Further, the use of synthetic wastewater as was the case in this study may give rise to uncertainty in wastewater characterization i.e. fractionation while following the standard ASM ratios. Changes in the feed solution configuration would affect the predicted bacterial concentration and consequently the relative contribution of microbial reactions to organic matter removal. Nevertheless the constituents of the synthetic wastewater were both constant and known making the application to the model easier.

6.3.1 Oxygen transfer

The extension of CWM1 to include physical re-aeration and root oxygen release from macrophytes among other processes has been recommended for the formulation of a full model for constructed wetlands (Langergraber et al. 2009). As oxygen consumption is very rapid in wastewater treatment technologies, and hence difficult to determine by direct measurement (Tyroller et al. 2010), indirect estimation of the oxygen transfer rate by both physical re-aeration and plant root oxygen release was attempted through inverse modeling. Direct oxygen transfer from air to water in the mesocosm is modeled in relation to an oxygen deficit as is usually considered in running waters. Low values obtained for the physical re-aeration rate could perhaps be explained by the fact that the air-water interface was subsurface and static, and neither the liquid nor gaseous phases were turbulently well mixed. However atmospheric oxygen diffusion is affected by a range of atmospheric conditions including temperature and relative humidity and wide range of oxygen diffusion, 0 to 28.6 g/m^2.d has been reported using different techniques and configurations of constructed wetlands (Wu et al. 2001).

Root oxygen release is a continuing subject of debate in subsurface flow CW research (Ojeda et al. 2008). Root oxygen release processes are said to depend on among others factors, plant species, plant biomass and season (Brix 1999; Caffrey and Kemp 1991), which could explain the difference in the oxygen release rates obtained between the three plant species and at the different incubation temperatures. Considering that the mesecoms had the same electron acceptor availability, the simulated root oxygen transfer at the lower temperatures potentially explains the observed greater COD removal in the *Carex* and *Schoenopletus* mesocoms but the absence of this temperature response in the *Typha* and the control mesocoms (Allen et al. 2002).While the root oxygen release rates determined this way (i.e. estimation by fitting measured data) are comparable to values obtained by other methodologies

(Table 6.9), it is generally recognized that wetland plants do not generate enough oxygen to fully remove pollutants from wastewater (Brisson and Chazarenc 2009; Brix 1997; Tanner 2001). Recent studies have shown that aerobic respiration plays only a minor role to the overall organic matter decomposition while anaerobic processes generally would play a major role in subsurface flow CWs with minimum oxygen renewal capacities (Llorens et al. 2011b; Ström et al. 2005). The depletion of SO_4^{-2}–S in both the simulation and measurements as shown in Fig. 6.3 also imply an insufficient oxygen supply to the columns because sulphate is reduced by the acetotrophic sulphate reducing bacteria only after oxygen and other more thermodynamically favorable electron acceptors have been depleted (Allen et al. 2002; Wiessner et al. 2005). Thus subsurface constructed wetlands with minimal oxygen renewal capacities such as horizontal subsurface flow wetlands should be designed as an anaerobic or an aerobic–anaerobic hybrid system rather than as an aerobic system (Bezbaruah and Zhang 2005).

Table 6.9. Plant oxygen release reported by various researchers

Plant	Amount of Oxygen release by plant	Remarks	Reference
Schoenoplectus	0.00104-0.00443 (g O_2 m^{-2}d^{-1}) 0.8 (g O_2 m^{-2}d^{-1})	Field and lab study Based on NH$_4^{-2}$ -N removal	Bezbaruah and Zhang (2005)
Typha	0.00384 -0.0064 (g O_2 Dry Weight^{-1} h^{-1})	Lab study	Jespersen et al. (1998)
Carex	0.64 ± 0.284(at 4°C)-0.67 ± 0.163(g O_2 m^{-2}d^{-1})	Microcosm	Tylor et al. (2009)

6.3.2   Influence of plants and temperature on bacterial biomass

The removal of wastewater constituents in CWM1 is associated with a specific microbial functional group, reflecting a fundamental characteristic of wastewater treatment facilities (including wetlands) where their functioning relies heavily on the metabolism of microorganisms contained within sludge or biofilm (Ragusa et al. 2004). Nevertheless microbial activity in constructed wetlands is still largely based on assumption and circumstantial evidence as there is a lack of effective indicators of biofilm function and health in water treatment systems (Faulwetter et al. 2009; Ragusa et al. 2004). By running the model, estimates of the initial concentration of the six bacteria group considered in CWM1 were made at different incubation temperatures. The sensitivity ranking (not shown) demonstrated that the initial bacteria concentration as boundary condition was significantly influential on the model processes. The estimated values for initial bacteria concentration (found to have a low to moderate uncertainty) in the modeled batch systems are comparable to and within the range of those determined from simulation studies of subsurface constructed wetlands with continuous flow e.g. Rousseau (2005) and Llorens et al. (2011). From the simulations, the planted mesocosm required a higher total bacteria concentration, compared with the unplanted controls, suggesting that incorporation of plant roots into substrate of constructed wetlands enhances microbial populations related to the transformation and degradation of pollutants in constructed wetlands.

The model agreed with the assumption that plants can affect bacterial density and activities in constructed wetland (Gagnon et al. 2007), with vegetation having a positive effect on the treatment efficiency for organics and nutrients such as nitrogen and phosphorus. Besides root surface area for microbial growth, it has been suggested that plant rhizosphere provides a source of carbon compounds through roots exudates and a micro-environment via root oxygen release that can affect microbial species composition and diversity (Vymazal 2011).

Temperature impact on microbiological process rates and on plant growth is expressed through growth and decay rates, as well as other kinetic parameters in this model. It is generally accepted that most bacteria activity will decrease at lower temperatures, which may in turn influence the wetland performance. However, both the measured and simulated results in this work demonstrate that, the resultant effect on the wetland performance may not necessarily be related to temperature. It was seen by running the model that the abundance of some bacterial function groups, such as heterotrophs is generally higher at lower temperatures compared to high temperatures. Low temperatures may cause population shifts by altering growth rates of the individual species in different ways changing the competitive situation between species (Kotsyurbenko 2005; Lew et al. 2004). The observed trend is in agreement with the results of Honda and Matsumoto (1983), who observed the growth capacity of a microbial film in a model trickling filter to increase as temperature fell. This is due to the autolysis coefficient which becomes lower at low temperatures (Honda and Matsumoto 1983). On the other hand calibration simulations indicated that methanogenic bacterial concentrations increased with temperature, apparently because in the model they are not limited by a substrate such as sulphate that was removed rapidly within the first 3 days of incubation. In this case sulphate deficiency leads to minimal utilization of acetate as an electron donor for the sulphate reducing bacteria, and acetate is mainly used by methanogenic bacteria (Kalyuzhnyi and Fedorovich 1998)

6.3.3   COD anaerobic degradation and sorption

Anaerobic transformation and degradation of COD was the dominant process (Fig.6.1) in agreement with other findings on microorganism activity in subsurface flow constructed wetlands (García et al. 2005; Imfeld et al. 2009; Llorens et al. 2011a). Considering that the redox potentials and sulfate concentrations were high immediately after filling the columns with fresh wastewater and generally decreased rapidly within 24hrs (Allen et al. 2002), organic matter degradation was likely achieved by anaerobic bacteria (Baptista et al. 2003; García et al. 2005). The widespread occurrence of anaerobic reactions in CWs has also been shown by Llorens et al. (2011b) with methanogenesis contributing 58–73% to the COD removal.

The greatest difference in performance (for COD) among the mesocosms was at lower temperature as reported in Stein and Hook (2005). At low temperatures the planted columns seem to have an enhanced capacity for COD adsorption coupled to a relatively active root oxygen release rate. Some previous studies have indicated that the measured behavior of wetland systems can only be modeled if COD adsorption is considered as an additional process (Henrichs et al. 2007). The model simulations of the current study indicate that the impact of adsorption-desorption processes is more significant at lower temperatures (12°C and 16°C) especially in the planted mesocosms, and that at these temperatures biochemical processes alone could not account for the observed COD removal (Fig. 6.2). This result underlines the importance of a COD adsorption-desorption process in batch operated constructed wetlands. Contribution by plant physical degradation and decay to increase of organic matter in the planted mesocosms was found to be insignificant within the 20 days simulation period.

6.3.4 NH$_4^+$–N removal mechanisms

Nitrogen transformation and removal mechanisms in constructed wetlands include mineralization (ammonification), ammonia volatilization, nitrification, denitrification, plant and microbial uptake, nitrogen fixation, nitrate reduction, anaerobic ammonia oxidation (ANAMMOX), adsorption, desorption, burial and leaching (Vymazal 2007). In this study the estimated contribution by plant uptake to the decline of NH$_4^+$–N concentration ranged between 0.1 % and 30% (Table 6.8). Plant uptake of nitrogen is typically seen as a less significant nitrogen removal mechanism (Tunçsiper 2009). Contribution by nitrification was also found to be low. This could have been influenced by the ammonia held and released by sorption process and the fact that available oxygen in the wastewater is quickly utilized by heterotrophic bacteria for the metabolism of organic carbon. This can also be seen from the results of the sensitivity analysis (Table 6.1) that shows sorption and heterotrophic bacteria rather than autotrophic bacteria growth rate and yield parameters as having the largest impact on simulated NH$_4^+$–N concentration. A high ammonia reduction rate in constructed wetlands with minimum oxygen renewal capacities is often less likely because the amount of oxygen available for bacterial oxidation of ammonium, or nitrification, is usually limited (Wu et al. 2001). The simulation highlights the significance of the sorption process for ammonium fate in the batch mesocosms, as also discussed in McBride and Tanner (1999). Effluent concentration of NH$_4^+$-N matched the measured concentration well only with the sorption process active in the simulation. This applied for the controls (unplanted columns) as well where an increase in NH$_4^+$–N due to organic nitrogen ammonification was predicted well.

6.3.5 Model advantages and limitations

The CWM1-AQUASIM (with adsorption and plant processes implemented) positively captured the influence of temperature, and the presence and species of macrophyte. The model elucidates the performance attributes and factors that describe the observed trends in COD, SO$_4^{-2}$-S and NH$_4^+$-N; differences in macrophyte oxygen transfer rates, density of microbial groups and temperature and the relative proportions of removal pathways of organic matter. Valuable insights into the parallel and interacting processes of a constructed wetland aid improved conceptualization and design of a constructed wetland system. The simplified batch hydraulic regime of the system modeled in this implementation of CWM1 in AQUASIM allowed a focus on the biokenetics of CWs. However, porous media hydrodynamics i.e. effect of dispersion, heterogeneity and dead zones must be considered for a realistic simulation and proper fit of the model to most operating CWs. For this, it is suggested that finite elements or finite difference models be used to describe water flow instead of the continuous stirred-tank reactor considered in this application (Langergraber et al. 2009). In pilot wetlands it is desirable that concentrations be tied to their locations, thereby creating the possibility of having aerobic, anoxic and anaerobic zones in the modeled wetland, as opposed to a "suspended culture". This can be achieved with use of biofilm models, with substrate conversion and bacteria growth combined with mass transport limitations into and within the biofilm matrix.

This model provides a promising tool for studying the dynamics of the key processes governing COD and nutrient dynamics in the wetland system, which is anticipated to promote an increased understanding of the performance aspects and sound conceptualization and design of constructed wetland systems. The CWM1 equations are based on commonly accepted Activated Sludge Models (ASMs), which enhances communication between wetland scientists as it introduces a kind of 'common language'. Due to the widespread application of the ASM models, literature provides much guidance for their stochiometric and kinetic parameters values as these models already have been applied in many case studies.

6.4   Conclusion

CWM1 was successfully implemented in AQUASIM software. The model was applied to data from unplanted and planted experimental batch operated constructed wetland mesocosms to simulate the transformation and degradation of COD, NH_4^+-N and $SO_4^{-2}-S$. Based on the model output the principal conclusions are:

(i)   The model was able to provide a reasonable fit to experimental data with the addition of physical oxygen re-aeration, plant and sorption processes.

(ii)   Sensitivity analysis showed the parameters with the highest sensitivities to be those related with micro-organisms kinetics and sorption processes.

(iii)   Plant root oxygen transfer rate varied with plant species and temperature. More oxygen was released to the mesocosms by plants at low temperature. Nevertheless, the oxygen transferred to the mesocosms by both plant root release and physical reaeration was insufficient compared with the oxygen demand of the wastewater.

(iv)   The better performance of the planted mesocosms was simulated well with a higher bacterial population than that of the unplanted mesocosms. Higher bacterial population was found at low temperature that high temperature.

(v)   COD degradation was mainly through anaerobic processes under these specific experimental conditions. SO_4^{-2}–S was used as an electron acceptor for COD removal and was depleted quickly at the beginning of incubations

(vii) NH_4^+–N was adsorbed on the gravel rapidly and further removed when the mesocosms were drained and the gravel exposed to the air. Nitrification and plant uptake of NH_4^+–N did not contribute significantly to the decline of NH_4^+–N in the experimental mesocosms

6.5   References

Allen, W. C., et al. (2002), 'Temperature and wetland plant species effects on wastewater treatment and root zone oxidation', Journal of Environmental Quality, 31 (3), 1010-16.

Baptista, J. D. C., et al. (2003), 'Microbial mechanisms of carbon removal in subsurface flow wetlands', Water Science and Technology, 48 (5), 127-34.

Bezbaruah, A. N. and Zhang, T. C. (2005), 'Quantification of oxygen release by bulrush (Scirpus validus) roots in a constructed treatment wetland', Biotechnology and Bioengineering, 89 (3), 308-18.

Brisson, J. and Chazarenc, F. (2009), 'Maximizing pollutant removal in constructed wetlands: Should we pay more attention to macrophyte species selection?', Science of the Total Environment, 407 (13), 3923-30.

Brix (1999), 'Functions of macrophytes in constructed wetlands.', Water Sci. Technol., 29 (4), 71–78.

Brix, H. (1997), 'Do macrophytes play a role in constructed treatment wetlands?', Water Science and Technology, 35 (5), 11-17.

Caffrey, J. M. and Kemp, W. M. (1991), 'Seasonal and spatial patterns of oxygen production, respiration and root-rhizome release in Potamogeton perfoliatus L. and Zostera marina L', Aquatic Botany, 40 (2), 109-28.

Faulwetter, Jennifer L., et al. (2009), 'Microbial processes influencing performance of treatment wetlands: A review', Ecological Engineering, 35 (6), 987-1004.

Gagnon, V., et al. (2007), 'Influence of macrophyte species on microbial density and activity in constructed wetlands', Water Science and Technology, 56 (3), 249-54.

García, Joan, et al. (2005), 'Effect of key design parameters on the efficiency of horizontal subsurface flow constructed wetlands', Ecological Engineering, 25 (4), 405-18.

Haberl, R. (1999), 'Constructed wetlands: A chance to solve wastewater problems in developing countries', Water Science and Technology, 40 (3), 11-17.

Henrichs, M., Langergraber, G., and Uhl, M. (2007), 'Modelling of organic matter degradation in constructed wetlands for treatment of combined sewer overflow', Science of the Total Environment, 380 (1-3), 196-209.

Honda, Y. and Matsumoto, J. (1983), 'The effect of temperature on the growth of microbial film in a model trickling filter', Water Research, 17 (4), 375-82.

Imfeld, Gwenaël, et al. (2009), 'Monitoring and assessing processes of organic chemicals removal in constructed wetlands', Chemosphere, 74 (3), 349-62.

Kadlec, Robert H. (2000), 'The inadequacy of first-order treatment wetland models', Ecological Engineering, 15 (1-2), 105-19.

Kalyuzhnyi, S. V. and Fedorovich, V. V. (1998), 'Mathematical modelling of competition between sulphate reduction and methanogenesis in anaerobic reactors', Bioresource Technology, 65 (3), 227-42.

Kivaisi, Amelia K. (2001), 'The potential for constructed wetlands for wastewater treatment and reuse in developing countries: a review', Ecological Engineering, 16 (4), 545-60.

Kotsyurbenko, O. R. (2005), 'Trophic interactions in the methanogenic microbial community of low-temperature terrestrial ecosystems', FEMS Microbiology Ecology, 53 (1), 3-13.

Kumar, J. L. G. and Zhao, Y. Q. (2011), 'A review on numerous modeling approaches for effective, economical and ecological treatment wetlands', Journal of Environmental Management, 92 (3), 400-06.

Langergraber, G. (2007), 'Simulation of the treatment performance of outdoor subsurface flow constructed wetlands in temperate climates', Science of the Total Environment, 380 (1-3), 210-19.

Langergraber, G. (2008), 'Modeling of processes in subsurface flow constructed wetlands: A review', Vadose Zone Journal, 7 (2), 830-42.

Langergraber, G., et al. (2009), 'CWM1: a general model to describe biokinetic processes in subsurface flow constructed wetlands', Water Science and Technology, 59 (9), 1687-97.

Lew, B., et al. (2004), 'UASB reactor for domestic wastewater treatment at low temperatures: a comparison between a classical UASB and hybrid UASB-filter reactor', Water Science and Technology, 49 (11-12), 295-301.

Llorens, E., Saaltink, M. W., and and Garcia, J. (2011a), 'CWM1 implementation in RetrasoCodeBright: First results using horizontal subsurface flow constructed wetland data', Chemical Engineering Journal, 166 (1), 224-32.

Llorens, E., et al. (2011b), 'Bacterial transformation and biodegradation processes simulation in horizontal subsurface flow constructed wetlands using CWM1-RETRASO', Bioresource Technology, 102 (2), 928-36.

McBride, Graham B. and Tanner, Chris C. (1999), 'Modelling biofilm nitrogen transformations in constructed wetland mesocosms with fluctuating water levels', Ecological Engineering, 14 (1-2), 93-106.

Meysman, Filip J. R., et al. (2003), 'Reactive transport in surface sediments. I. Model complexity and software quality', Computers & Geosciences, 29 (3), 291-300.

Moutsopoulos, Konstantinos N., et al. (2011), 'Simulation of hydrodynamics and nitrogen transformation processes in HSF constructed wetlands and porous media using the advection-dispersion-reaction equation with linear sink-source terms', Ecological Engineering, 37 (9), 1407-15.

Ojeda, E., et al. (2008), 'Evaluation of relative importance of different microbial reactions on organic matter removal in horizontal subsurface-flow constructed wetlands using a 2D simulation model', Ecological Engineering, 34 (1), 65-75.

Puigagut, Jaume, et al. (2007), 'Subsurface-flow constructed wetlands in Spain for the sanitation of small communities: A comparative study', Ecological Engineering, 30 (4), 312-19.

Ragusa, S. R., et al. (2004), 'Indicators of biofilm development and activity in constructed wetlands microcosms', Water Research, 38 (12), 2865-73.

Reichert (1998), 'AQUASIM 2.0 - User Manual Computer Program for the Identi cation and Simulation of Aquatic Systems', Swiss Federal Institute for Environmental Science and Technology (EAWAG) CH - 8600 Dubendorf Switzerland.

Reichert, Peter (1995), 'Design techniques of a computer program for the identification of processes and the simulation of water quality in aquatic systems', Environmental Software, 10 (3), 199-210.

Rousseau, D. P. L. (2005), 'Performance of constructed treatment wetlands: model-based evaluation and impact of operation and maintenance. PhD Thesis, Ghent University, Ghent, Belgium (available from http://biomath.ugent.be/publications/download/).'.

Rousseau, D. P. L., Vanrolleghem, P. A., and De Pauw, N. (2004), 'Model-based design of horizontal subsurface flow constructed treatment wetlands: a review', Water Research, 38 (6), 1484-93.

Sanchez, E., et al. (2004), 'Changes in biomass, enzymatic activity and protein concentration in roots and leaves of green bean plants (Phaseolus vulgaris L. cv. Strike) under high NH4NO3 application rates', Scientia Horticulturae, 99 (3-4), 237-48.

Solano, M. L., Soriano, P., and Ciria, M. P. (2004), 'Constructed Wetlands as a Sustainable Solution for Wastewater Treatment in Small Villages', Biosystems Engineering, 87 (1), 109-18.

Stein, Otto R., et al. (2007), 'Seasonal influence on sulfate reduction and zinc sequestration in subsurface treatment wetlands', Water Research, 41 (15), 3440-48.

Ström, Lena, Mastepanov, Mikhail, and Christensen, Torben (2005), 'Species-specific Effects of Vascular Plants on Carbon Turnover and Methane Emissions from Wetlands', Biogeochemistry, 75 (1), 65-82.

Tanner (2001), 'Plants as ecosystem engineers in subsurface-flow treatment wetlands', Water Science and Technology, 44 (11-12), 9-17.

Tunçsiper, B. (2009), 'Nitrogen removal in a combined vertical and horizontal subsurface-flow constructed wetland system', Desalination, 247 (1-3), 466-75.

Tyroller, Lina, et al. (2010), 'Application of the gas tracer method for measuring oxygen transfer rates in subsurface flow constructed wetlands', Water Research, 44 (14), 4217-25.

Vymazal (2007), 'Removal of nutrients in various types of constructed wetlands', Science of the Total Environment, 380 (1-3), 48-65.

Vymazal, J. (2011), 'Plants used in constructed wetlands with horizontal subsurface flow: a review', Hydrobiologia, 1-24.

Wiessner, A., et al. (2005), 'Sulphate reduction and the removal of carbon and ammonia in a laboratory-scale constructed wetland', Water Research, 39 (19), 4643-50.

Wu, M. Y., Franz, E. H., and Chen, S. (2001), 'Oxygen fluxes and ammonia removal efficiencies in constructed treatment wetlands', Water Environment Research, 73 (6), 661-66.

Specific Appendix

Table 1. Mathematical equations for plant related process, physical re-aeration, adsorption-desorption processes as implemented in AQUASIM alongside the CWM1

Process	Process rate
Plant Growth NH$_4^+$–N	$\dfrac{1}{d \times eps} \times K_{pl} \times \dfrac{S\_NH}{K_{pnh} + S\_NH} \times X_{pi}$
Plant Growth NO$_3^-$-N	$\dfrac{1}{d \times eps} \times K_{pl} \times \dfrac{S\_NO}{K_{pno} + S\_NO} \times \dfrac{K_{pnh}}{K_{pnh} + S\_NH} \times X_{pi}$
Plant Degradation	$\dfrac{1}{d \times eps} \times K_{\deg rad} \times X_{pd}$
Plant Decay	$\dfrac{1}{d \times eps} \times b_p \times X_{pi}$
Plant oxygen leaching rate	$\dfrac{K_{rol}}{d \times eps} \times \exp\big((S\_O_{sat} - S\_O) - 1\big)$
Physical reaeration	$K_{la} \times (S\_O_{sat} - S\_O)$
COD Adsorption	$\omega \times \big(k\_S\_COD \times X\_S - X\_COD_{Ad}\big)$
Ammonia Adsorption	$alpha * \left(S\_NH - (\dfrac{X\_NH}{PartitionCoefficient})^{\frac{1}{m}} \right)$

Table 2. Parameter and variable description and value

Parameter	Description	Unit	Value
d	Rooting depth	m	0.5
eps	Matrix material porosity as fraction		0.27
K_{pl}	Plant relative growth rate, function of season	1/d	0.033
K_{pnh}	Ammonium half-saturation coefficient for plant growth	gNH_4^+-N/m^3	0.3
X_{pi}	Living plant biomass	$gCODplant/m^2$	0 - 300
K_{pno}	Nitrate half-saturation coefficient for plant growth	gNO_3^--N/m^3	0.1
$K_{deg\,rad}$	First order plant physical degradation constant	1/d	0.01
X_{pd}	Dead standing plant biomass	$gCODplant/m^2$	0 - 300
b_p	Decay coefficient for living plant material, function of season		0.002
K_{rol}	Root oxygen loss parameter as per day	m/day	
S_O_{sat}	Oxygen saturation concentration	mgO_2/l	
K_{la}	Oxygen reaeration coefficient	1/d	
ω	First-order exchange rate	1/h	0.1-1
k_S_COD	Empirical coefficient for COD adsorption	$[mg_{COD}\cdot kg_{substrate}^{-1}]$	1-3
X_COD_{Ad}	Concentration of sorbed COD	$gCOD/m^3$	
$alpha$	Specific sorption rate coefficient	1/day	2
$PartitionCoefficient$	Solid-liquid (ammonium) partition coefficient	l/kg gravel	3
m	Freundlich isotherm exponent		0.5

Chapter 7: Simulation of batch-operated experimental wetland mesocosms in AQUASIM biofilm reactor compartment

This chapter has been submitted for publication at the Journal of Environmental Management as: Simulation of batch-operated experimental wetland mesocosms in AQUASIM biofilm reactor compartment. Njenga Mburu, Diederik P.L. Rousseau, Otto R. Stein, Piet N.L. Lens. August 2013

Abstract

In this study, a mathematical biofilm reactor model based on the structure of the Constructed Wetland Model No.1 (CWM1) coupled to AQUASIM's biofilm reactor compartment has been used to reproduce the sequence of transformation and degradation of organic matter, nitrogen and sulphur observed in a set of constructed wetland mesocosms and to elucidate the biofilm growth dynamics of a multispecies bacterial-biofilm in a subsurface constructed wetland. Experimental data from 16 wetland mesocosms operated under greenhouse conditions, planted with three different plant species (*Typha latifolia, Carex rostrata, Schoenoplectus acutus*) and an unplanted control were used in the calibration of this mechanistic model. Within the mesocosm, a thin (predominantly anaerobic) biofilm was simulated with an initial thickness of 49 μm (average) and in which no concentration gradients developed. The biofilm density and area, and the distribution of the microbial species within the biofilm were evaluated to be the most sensitive biofilm properties; while the substrate diffusion limitations were not significantly sensitive to influence the bulk volume concentrations. The simulated biofilm density ranging between 105,000 to 153,000 gCOD/m^3 in the mesocosms was observed to vary with temperature, the presence of and the species of macrophyte. The biofilm modelling was found to be a better tool than the suspended bacterial modeling approach to show the influence of the rhizosphere configuration on the performance of the constructed wetlands.

7.0   Introduction

Subsurface flow constructed wetlands (SSF-CWs) are finding extensive application for domestic and municipal wastewater treatment (Haberl, 1999; Neralla et al., 2000; Vymazal, 2010; Mburu et al., 2013b) because of their simple and robust configuration together with low energy requirements and operating cost. SSF-CWs are generally constructed with a porous material (e.g. soil, sand, or gravel) as a substrate for growth of rooted wetland plants in addition to various microbes. The microorganisms and their extracellular products adhere to the solid support provided by the porous media and plant roots, forming a biofilm layer in which the contaminant compounds disperse and are degraded by the microorganisms (Wichern et al., 2008; Kadlec and Wallace, 2009). Thus, the organic content in the wastewater is reduced by biological degradation rather than by simple screening (Krasnits et al., 2009). Aerobic respiration, denitrification, sulphate reduction and methanogenesis are the principal biochemical reactions involved in the oxidation and net removal of organic matter in subsurface flow constructed wetland systems (Baptista et al., 2003; Caselles-Osorio et al., 2007; Langergraber et al., 2009). Transformation processes also include abiotic chemical reactions, such as adsorption of a solute onto the biomass or the solid porous material. Thus SSF-CWs clearly behave as complex reactors (Ojeda et al., 2008). Accordingly, in order to evaluate such a system, mechanistic mathematical models could be very helpful to facilitate the interpretation and quantification of the ongoing biogeochemical processes.

Despite there is a recognition that the improvement of water quality in treatment wetland applications is primarily due to microbial activity (Faulwetter et al., 2009; Kadlec and Wallace, 2009), the mechanistic understanding of the dynamics of microbial biofilm biomass, activity, and community composition in constructed wetlands is still evolving, albeit significantly in the past couple of years (Faulwetter et al., 2009; Truu et al., 2009; Samsó and Garcia, 2013). These aspects have been completely overlooked in traditional wetland models using reaction rate constants (Rousseau et al., 2004). Only recently constructed wetland mechanistic modeling started to incorporate these aspects (Langergraber, 2001; Langergraber and Šimůnek, 2005; Rousseau, 2005). However, barely a few of the mechanistic models consider multispecies bacterial biofilms (McBride and Tanner, 1999; Langergraber and Šimůnek, 2005; Mayo and Bigambo, 2005; Langergraber and Šimůnek, 2012; Samsó and Garcia, 2013), but rather are formulated with "suspended cells" (i.e. bacterial community without substratum) under batch or continuous flow modes (Wynn and Liehr, 2001; Mayo and Bigambo, 2005; Rousseau, 2005; Llorens et al., 2011b; Mburu et al., 2012a). Further, there are variations in the modeling of microbial reactions including type and number of bacterial populations considered and kinetics of growth and processes affecting the bacterial-biofilm growth (Kumar and Zhao, 2011) due to the complex nature of constructed wetland systems. In this context, the biokinetic Constructed Wetland Model number 1 (CWM1) (Langergraber et al., 2009) is seen as the most advanced theoretical biokinetic model developed for SSF-CWs.

CWM1 has been implemented in different simulation platforms (Llorens et al., 2011a; Langergraber and Šimůnek, 2012; Mburu et al., 2012b; Mburu et al., 2013a; Samsó and Garcia, 2013) and the resulting codes have been used to match experimentally measured effluent pollutant concentrations. Although these models provide insights into the behavior of the SSF constructed wetland, they neglect certain potentially important phenomena influencing microbial reactions, such as diffusion limitation or the stratification of metabolic processes in the biofilm when several populations of bacteria are present, or the possible influence of macrophyte type on the development of biofilm biomass (Gagnon et al., 2007; Zhang et al., 2010). Knowledge on the growth dynamics of bacterial biofilms is essential for the design conceptualization of treatment processes in constructed wetlands. Compared with reactors with suspended bacteria, fixed biofilm bacteria reactors can be operated at high biomass concentrations in the reactor. This implies that biofilm units often require less land area than suspended bacteria units (Wik, 1999). Thus, the attached bacteria-biofilm modeling approach should give better system insights to help to improve the performance of constructed wetlands by providing a scientific basis to find the optimal design and operating conditions of constructed wetlands systems (Rousseau, 2005).

A biofilm modeling approach has been used here to reproduce the sequence of transformation and degradation of organic matter, nitrogen and sulphur observed in a set of constructed wetland mesocosms and to elucidate the biofilm growth dynamics in a multispecies bacterial-biofilm. The growth of six microbial groups (heterotrophic, autotrophic nitrifying, fermenting, acetotrophic methanogenic, acetotrophic sulphate reducing and the sulphide oxidising bacteria) and the subsequent consumption of electron donors and acceptors in 16 batch operated subsurface flow wetland mesocosms operated under controlled greenhouse conditions with three different plant species (*Typha latifolia, Carex rostrata, Schoenoplectus acutus*) and an unplanted control is simulated. The processes occurring in the biofilms attached to the gravel and plant roots in the mesocosms are simulated in the one-dimensional (1-D) mathematical biofilm model of the simulator AQUASIM (Reichert, 1998), a programme for identification and simulation of aquatic systems. For the rate equations and kinetics of the microbiological processes, the Constructed Wetland Model No.1

(CWM1) biokinetic model as described in Langergraber et al. (2009) is used. The results are compared and contrasted with those in the work of Mburu et al. (2012) in which a non-biofilm (i.e. suspended) multipopulation bacterial growth and uptake approach together with plant related processes (growth, physical degradation, decay, and oxygen leaching), physical re-aeration, as well as adsorption and desorption processes for COD and ammonium were applied in the simulation of the 16 mesocosms to describe the transformation and degradation processes of organic matter, nitrogen and sulphur (Mburu et al., 2012a). Thus although biofilm modeling may represent a theoretical improvement over the "suspended bacteria" approach in constructed wetland modeling, a direct comparison of the two modeling approaches is important to determine if the biofilm modeling approach yields new qualitative information, specifically on the possible influence of the boundary conditions of temperature and macrophyte species on the prediction of substrate removal in constructed wetlands.

7.1   Methodology

7.1.1   The experimental constructed wetlands

The experimental constructed wetlands were operated under controlled greenhouse conditions at Montana State University in Bozeman (Montana, USA). Details of column design, construction and planting, as well as sampling and measurement are described in Allen et al. (2002) and Stein et al. (2006). Briefly, 16 subsurface constructed wetland mesocosms were constructed from polyvinyl chloride (PVC) pipes (60 cm in height × 20 cm in diameter) and filled to a depth of 50 cm with washed pea-gravel (0.3–1.3 cm in diameter). Four columns each were planted with *Carex rostrata* (Northwest Territory sedge), *Schoenoplectus acutus* (hardstem bulrush) and *Typha latifolia* (broadleaf cattail), while four were left unplanted as controls. A series of 3-6-9-20-d incubations with synthetic wastewater was conducted over 20 months at temperatures ranging from 12 to 24°C at 4°C steps. A synthetic wastewater simulating secondary domestic effluent was used with mean influent concentrations of 470 mg/l COD, 44 mg/l N (27 Org-N, 17 NH_4^+–N), 8 mg/l PO_4^{3-}–P, and 14 mg/l SO_4^{2-}–S. Columns were gravity drained 3 days prior to each incubation and then again at the start of each incubation. Upon each emptying, columns were refilled from above with new wastewater. Sampling from all 16 columns occurred at days 0, 1, 3, 6, 9, 14 and 20 of each incubation and those sub-samples were analyzed afterwards for the constituents.

7.1.2 Model description and implementation

The constructed wetland mesocosm is mathematically described as a reactor with completely mixed bulk water volume and with a biofilm growing on a substratum (gravel media and plant roots) surface inside the reactor. The mesocosm was implemented into the biofilm compartment of the AQUASIM 2.1d (win/ mfc) software (Reichert, 1998). The biofilm reactor compartment in AQUASIM enables the simulation of biofilm systems with several microbial species and substrates. It describes the spatial distribution and development in time of dissolved and particulate components in the biofilm, as well as the development in time of the biofilm thickness. The biofilm is divided into a liquid phase consisting of water (80%) in which the dissolved substances are transported by diffusion and a solid matrix (20%) consisting of particulate components such as active and inactive bacteria and their extracellular polymeric substances (EPS). The compartment was configured as:

- "Unconfined" reactor (i.e. volume of bulk liquid was assumed constant and did not change with biofilm thickness, whereas the biofilm can grow freely as may be the case in a trickling filter (Wanner and Morgenroth, 2004)).
- A rigid structure (there is no diffusive mass transport of solids i.e., the biofilm matrix can change its volume due to microbial growth and decay only).
- No suspended solids in the pore volume (no particle transport through the biofilm).
- No surface or volume attachment or detachment (there is no exchange of particulate components between the biofilm solid matrix and the bulk liquid).
- Diffusivities were taken from Boltz et al. (2011), considering that the diffusivity of a solute inside the biofilm is generally lower than that in water because of the tortuosity of the pores and minimal biofilm permeability.

The transformation and degradation processes were defined based on the bio-kinetic model CWM1. The values of the kinetic and stoichiometric parameters required by the model are available in Langergraber et al. (2009), with the exception of those that were modified in Mburu et al. (2012). Dynamic processes were used for the growth and decay rate of all bacterial groups as included in the CWM1 model. Because of the serial execution of the equation of the heterotrophic bacteria processes as described in CWM1, growth failed when switching between readily biodegradable COD and the fermentation product acetate, both under aerobic and anoxic conditions. The implementation of these process rates as described in Langergraber et al. (2009) was thus not possible. Instead, the implementation was carried out as described in the work of Llorens et al. (2011), where the solution was to divide the heterotrophic bacteria group into two subgroups according to the substrate they consume.

The biofilm is modeled as a film growing on the spherical surfaces of the gravel media inside the mesocosm and assumed to be an ideal biofilm of uniform thickness (L_F) and density (*rho*), with the diffusion and kinetic coefficients assumed to be constant throughout the wetland unit. The equations to calculate the biofilm growth area on the washed pea gravel followed the approach of Wichern et al. (2008). In an idealized way, the washed gravel is approximated by spheres which all have the same diameter and which can touch each other at up to eight points. Where the spheres touch, biomass growth on the surface of the spheres is not possible. Thus, it is possible to determine the loss of biofilm surface area (A_{loss}) between two pieces of gravel (considered as spheres) in relation to the radius (r) of the single sphere and the thickness of the biofilm (L_F) with the equation:

$$A_{loss} = B\pi L_F (2r + 2L_F) \; \left[m^2 \right] \tag{7.1}$$

where B represents the number of contact points per sphere. The number of pieces of gravel (N) in the mesocosms volume (V) is estimated depending on the porosity ε as:

$$N = \frac{(1-\varepsilon)V}{\dfrac{\pi}{6}(2r)^3} \; [-] \tag{7.2}$$

The remaining surface area of the biofilm $A_{remaining}$ in relation to the number of contact points, the diameter of the spheres, and the biofilm thickness amounts to:

$$A_{remaining} = N\pi(2r)^2 - NA_{loss} \; \left[m^2 \right] \tag{7.3}$$

Reduction in concentration of any substrate is modeled as a mass-transfer or boundary layer mass transfer resistance, R_L ($=L_L/D$), which depends on the diffusivity (D) of a substrate inside the biofilm and an accurate estimate of the mass transfer liquid layer of thickness L_L. Very low diffusion coefficients will lead to high boundary layer resistance and the model will not yield a converging solution. The diffusion coefficients of the components in the biofilm were taken as their "effective" values approximated as 0.8 times their diffusion coefficient through pure water (Boltz et al., 2011). Temperature dependency of diffusion coefficients was accounted for according to:

$$D(T) = D(20^{o}C) . \frac{273 + T}{273 + 20^{o}C} . \frac{\mu(20^{o}C)}{\mu(T)}, \; [m^{2}d^{-1}] \qquad (7.4)$$

where D is the diffusion coefficient, T the temperature in ^{o}C, and μ the dynamic viscosity of water in $N\,m^{-2}s$ (Boltz et al., 2011).

Five plant processes (growth, physical degradation, decay, and oxygen leaching), physical re-aeration, as well as adsorption and desorption processes for COD and ammonium nitrogen were also included as dynamic processes following the work of Mburu et al. (2012).

To run the model, 16 inputs characterizing the influent (Oxygen, COD fractions, N compounds, and S compounds), and one input for water temperature are required. Other inputs concerning initial data i.e. the biofilm (density, thickness, area and volume fractions), the boundary liquid layer thickness, the reactor volume and the diffusion coefficients of 13 substrate is necessary. Fractionation of the influent wastewater COD was based on standard ratios given in the ASM models (Henze et al., 2000) and implemented following the work of Mburu et al. (2012).

7.1.3 Sensitivity analysis

A sensitivity analysis was carried out to recognize the most important parameters influencing the prediction of carbon, nitrogen and sulphur concentrations and the development of the biofilm. With the "sensitivity analysis" function in AQUASIM, it is possible to investigate whether the time series of the calculated values are affected noticeably by a change in the value of a model parameter. The sensitivity analysis feature enables calculation of linear sensitivity functions of arbitrary variables with respect to each of the parameters included in the analysis (Reichert, 1995). The sensitivity analysis results described in this study are those of the absolute-relative sensitivity function of AQUASIM (Eq. 7.5) that computes the absolute change in a model output variable, y, for a 100% change in any parameter of interest, p:

$$\delta_{y,p}^{a,r} = p \frac{\partial y}{\partial p} \tag{7.5}$$

This makes quantitative comparisons of the different parameters on a common variable possible.

The uncertainty is determined by using the error propagation formula (Eq. 7.6), which is based on the linearized propagation of standard deviations of the parameters of interest, neglecting their correlation:

$$\sigma_y = \sqrt{\sum_{i=1}^{m} \left(\frac{\partial y}{\partial p} \right)^2 \sigma^2_{p_i}} \tag{7.6}$$

Where p_i are the uncertain model parameters, σ_{p_i} their standard deviations, $y(p_i \cdots\cdots p_m)$ the solution of the model equations for a given variable at a given location and time, and σ_y is the approximate standard deviation of the model result. Identifiability of the model parameters was evaluated by use of the parameter correlation matrix in AQUASIM.

7.1.4   Model calibration and simulations

Data based on bulk measurements of COD, NH_4^+–N and SO_4^{2-}–S at $12°C$, $16°C$, $20°C$ and $24°C$ from the unplanted (control) mesocosms were used for optimization of the microbial biokinetic parameters of CWM1 and the calibration of the biofilm parameters. The physical re-aeration coefficient, initial amount of sorbed ammonia and the COD adsorption parameters were adopted as determined in the work of Mburu et al. (2012). To optimize the parameter sets, the result of the sensitivity analysis was used to guide the selection and calibration of the kinetic coefficients and the biofilm parameters with the "parameter estimation" function of AQUASIM. The function attempts to determine unknown values of model parameters by iteratively best-fit matching time-series of calculated and measured values. The simulations were conducted with data from planted mesocosm, at $12°C$, $16°C$, $20°C$ and $24°C$.

7.2   Results

7.2.1   Sensitivity and identifiability analysis

Prior to calibration of the biofilm parameters, the parametric sensitivity of the dynamic model was conducted in AQUASIM (Table 7.1). The importance of the constructed wetland biofilm structure was reflected in the dependence of the state variables on the biofilm density and area. Table 7.1 shows the sensitivities of the main bulk liquid concentrations (i.e. COD, NH_4^+-N and SO_4^{2-}-S) on biofilm characteristics. Here, "strong effect", "significant", "moderate" and "insignificant" indicate $SF \geq 1$, $1 > SF \geq 0.1$, $0.1 > SF \geq 0.01$ and $SF < 0.01$, respectively, where SF is the absolute-relative sensitivity function (unit of g-COD/m^3, g-N/m^3 or g-S/m^3).

Table 7.1. Sensitivity of the biofilm parameters on the bulk liquid concentrations

	Rho	D	LL	LF_{ini}	A	eps_X
COD	+++	-	++	++	+++	++
NH_4^+-N	+++	-	+	++	++	++
SO_4^{2-}-S	+++	-	+	++	++	++

(-) insignificant effect; (+) moderate effect; (++) significant effect; (+++) strong effect; (rho: biofilm density, D:substrate diffusivity, LL: liquid layer thickness, LF_{ini} initial biofilm thickness, A:area of biofilm, eps_X: biomass volume fraction)

The correlation matrix for the biofilm parameters (Table 7.2) shows the biofilm density and the biofilm area to be strongly correlated and hence not simultaneously theoretically identifiable from the measured data (Petersen et al., 2001). This suggests that other factors (e.g. diffusivity or mass transport limitation within the biofilm) may have influenced parameter identifiability, especially for parameters that are otherwise uncorrelated or with a low linear dependency (Brockmann et al., 2008).

Table 7.2. Correlation matrix during biofilm parameter estimation in AQUASIM

	A	LFini	LL	rho
A	1			
LFini	-7.0E-06	1		
LL	-7.9E-06	-4.5E-07	1	
rho	1	1.7E-01	0.46	1

7.2.2   Biofilm properties

7.2.2.1   Simulated volume fractions and the activity of microbial species

The simulated volume fractions of the microbial species within the biofilm at different incubation temperatures are presented in Table 7.3. The fractions represent the interactions within the biofilm depth among the bacteria species involved and their competition for existence within the biofilm as a function of the substrate flux. The simulated volume fractions were not varying significantly with temperature, while the sulphide oxidising bacteria were not growing in the biofilm probably due to oxygen and nitrate limitation during the biofilm development.

Table 7.3. Simulated volume fractions of bacterial functional groups of CWM1 in the control mesocosms

Temp °C	Bacterial Functional group					
	X_FB	X_ASRB	X_AMB	X_A	X_H	X_SOB
12	0.026	0.047	0.057	0.002	0.047	0.000
16	0.029	0.047	0.061	0.005	0.038	0.000
20	0.029	0.047	0.061	0.003	0.039	0.000
24	0.029	0.048	0.062	0.005	0.037	0.000

X_FB: Fermenting bacteria; X_ASRB: Acetotrophic sulphate reducing bacteria; X_AMB: Acetotrophic methanogenic bacteria; X_A: Autotrophic nitrifying bacteria; X_H: Heterotrophic bacteria; X_SOB: Sulphide oxidising bacteria

7.2.2.2 Physical and geometric parameters

The characterization of the physical and geometrical parameters of the biofilm in the control and the planted mesocosms is presented in Tables 7.4 and 7.5, respectively. The calibrated values for initial biofilm thickness (LF_{ini}), biofilm density (rho), liquid boundary layer (LL) and the estimated biofilm area (A) were obtained by fitting the model to the experimental measurements of the bulk liquid concentrations of the wetland mesocosms set-ups observed at the different incubation temperatures.

Table 7.4. Calibrated values of the initial biofilm thickness (LF_{ini}), density (rho), liquid layer (LL) and area (A) in the control mesocosms

Temp °C	Parameter			
	LF_{ini} (m)	rho (g/m^3)	LL (m)	A (m^2)
12	4.74E-05	134025	6.24E-05	8.42
16	3.84E-05	132655	6.29E-05	8.41
20	6.33E-05	105190	5.93E-05	8.4
24	4.74E-05	108512	3.40E-05	8.43

Table 7.5. Calibrated values of the initial biofilm thickness density *(rho)*, and area *(A)* in the planted mesocosms

Set-up	Temp °C	Parameter *rho* (g/m³)	*A* (m²)
	12	152519	8.55
Carex	16	139877	8.58
	20	143015	8.57
	24	127684	8.57
	12	150290	8.57
Schoenoplectus	16	143926	8.58
	20	144135	8.58
	24	143189	8.58
	12	148266	8.57
Typha	16	145365	8.55
	20	137446	8.55
	24	130818	8.57

The initial biofilm thickness (average 49 μm) was observed to shrink with time during the batch simulations without reaching a steady state under the experimental conditions, as shown in Figure 7.1. It is clear that the progressive decrease in biofilm thickness follows a trend of diminishing substrate availability under the batch loading conditions. The thickness of the biofilm is governed by the flux of substrate to the biomass, as well as the growth and decay of the micro-organisms in the model. The attachment and detachment of cells at the biofilm surface and inside the biofilm was not considered in this work.

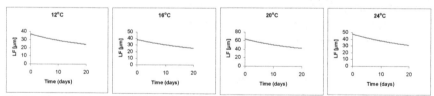

Fig. 7.1. Simulated development of the biofilm thickness in the control mesocosms

7.2.3 Bulk volume simulations

Model predictions showed a good qualitative agreement with the measured bulk volume concentrations of COD, NH_4^+-N and SO_4^{2-}-S (Figures, 7.2, 7.3 and 7.4). However, discrepancies exist between measured and simulated NH_4^+–N and SO_4^{2-}–S concentrations. The simulations of NH_4^+–N and SO_4^{2-}–S conversion/consumption show an offset (both over-estimation and under-estimation) in the planted mesocosms. These discrepancies between the measured and simulated values might be due to the fact that the mean density of the biofilm in the model is estimated rather than experimentally determined. Ideally, the mean biofilm density is known and is considered as an input parameter. Alternatively omitting the the influence of plant roots on the biofilm development represents an oversimplification (of the mathematical model) in attempting to determine biofilm model parameters based on experimental data. This is possibly the case, as the obtained planted-unplanted biofilm density ratio was found to range between 7 to 13 as described in Munch et al. (2005). Further, the non-development of the sulphide oxidising bacteria in the biofilm (Table 7.3) caused the poor fit for the SO_4^{2-}–S profiles.

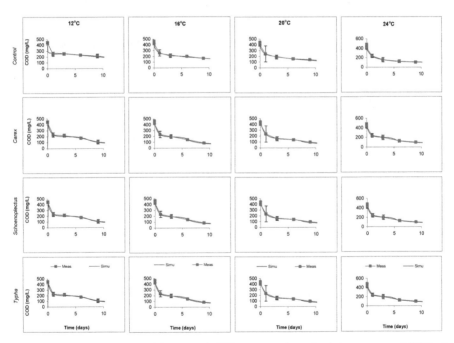

Fig. 7.2. Simulated compared with measured COD concentrations. Symbols are means (±S.D.) of observed concentrations from four replicates for each treatment.

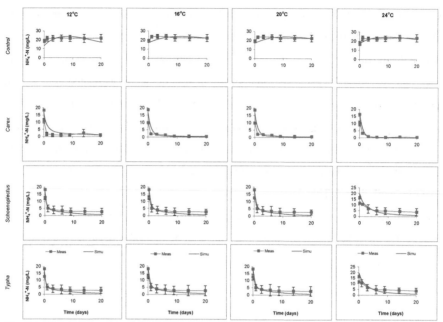

Fig. 7.3. Simulated compared with measured NH$_4^+$-N concentrations. Symbols are means (±S.D.) of observed concentrations from four replicates for each treatment.

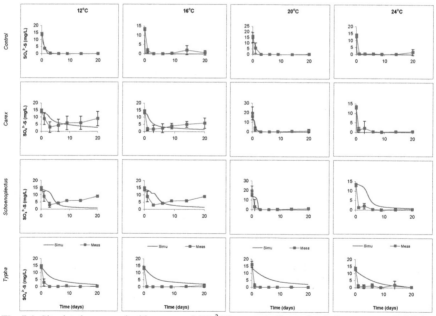

Fig. 7.4. Simulated compared with measured SO$_4^{2-}$-S concentrations.. Symbols are means (±S.D.) of observed concentrations from four replicates for each treatment.

7.3   Discussion

7.3.1   Biofilm versus Suspended bacterial growth modeling approach for simulating constructed wetland performance

The bacterial biofilm modeling approach is more complex than the suspended bacterial approach presented in Mburu et al. (2012), because the reactor mass balance is coupled with a diffusion reaction equation for the substrate in the biofilm. The biofilm model must account for the spatial aspects of biofilms, most notably the distribution of bacteria and substrates across the depth of a biofilm (Reichert, 1998).

Substrate gradients in the biofilm as a consequence of diffusion and reaction were not observed. The growth of bacterial biomass through substrate consumption is essentially under the same conditions or behavior as for the suspended bacterial model. Nevertheless, the evaluated bacterial volume fractions (Table 7.3), biofilm density and area (Table 7.4) show it was possible to evaluate more clearly how the individual functional groups developed in the biofilm of the constructed wetland. For example, the reason for the better performance of the planted mesocosm (with respect to substrate removal) compared to the unplanted mesocosm, is simulated as an increased colonizable surface area on which the biofilms can grow, together with the concomitant increase in biofilm density (due to incorporation of the plant). This is not obvious from the suspended bacteria model in Mburu et al. (2012), where these individual components of bacterial volume fractions, biofilm area and density are all lumped together as microbial concentration. Further, the biofilm density among the planted mesocosms is varying rather more significantly than the area of the biofilm, suggesting that the biofilm density (which is in qualitative agreement with the bacterial concentrations of the suspended biomass model) was the more significant factor influencing the performance of the wetlands. Hence, the dependence of the substrate removal on the rhizosphere configuration, i.e. the extra microbial attachment surface area and the potential of enhanced biofilm density provided by different macrophyte species strongly determines the constructed wetland performance.

7.3.2   Influence of macrophyte species and temperature on the biofilm density

Biofilm density and thickness are the main design parameters used to evaluate the substrate consumption rate in biofilms (Vanhooren, 2002). In this study, the biofilm density was observed to vary with the presence and species of macrophyte as well as the temperature. Planted mesocosms were found to develop a higher density biofilm compared with the unplanted mesocosms. The presence of plants enhanced the microbial density and activity in experimental microcosm studies as has been reported by Gagnon et al. (2007) with results showing a bacterial density ratio of 10.3 between planted and unplanted microcosms. The ratio obtained from the simulations is, however, much lower, suggesting the simplifying assumptions made in the mathematical formulation of the biofilm with regard to microorganisms-plant interactions may cause some disagreements between the model output and field observations of the planted versus unplanted wetlands. For example, the estimation of the biofilm area (Table 7.2 and 7.3) is based on the surface area provided by the gravel (equation 7.3), ignoring the effect of plant root development. It is generally assumed that planted wetlands outperform unplanted controls, mainly because the plant rhizosphere stimulates microbial communities either through high carbon availability in the rhizosphere resulting of root exudates or extra attachment sites of the root surface correlated to plant species root morphology and development (Munch et al., 2005; Gagnon et al., 2007).

Biofilms are known to vary in their density (Lazarova and Manem, 1995). Biofilms with densities from 10 to 130 kg dry mass/m^3 wet volume have been reported in different aquatic systems including aquifers and wastewater treatment systems (Zysset et al., 1994; McBride and Tanner, 1999; Vanhooren, 2002; Melo, 2005). The simulated biofilm densities lie in this range, considering bacterial concentrations may be converted from COD units to DM units by using the conversion factor of 1.222 gCOD (g biomass)$^{-1}$, as proposed in Rousseau (2005). There are many factors that could be responsible for the variation in biofilm density, such as culture morphology, i.e. changes in species, and amount of inactive material, and changes in biofilm porosity or lysis (Şeker et al., 1995; Wik, 1999). The simulated biofilm density variation with respect to temperature in this study (Table 7.5) is in agreement with the results of Honda and Matsumoto (1983), who observed the growth capacity of a

156

microbial film in a model trickling filter to increase as temperature fell. This is due to the autolysis coefficient which becomes lower at low temperatures (Honda and Matsumoto, 1983).

7.3.3 Sensitivity of the biofilm parameters

The biofilm density, area and microbial volume fractions were the most sensitive biofilm characteristics for the majority of the variables in the bulk liquid zone (Table 7.1). This is an indication that the total biomass concentration developed, the flux area available for diffusion of substrate as well as the bacterial composition of the biofilm are important parameters in the dynamic simulation of the wetland mesocosms.

The sensitivity analysis revealed the diffusion coefficients of components into the biofilm have a low or insignificant sensitivity to the final simulation results at all temperatures. It appears that pore water diffusion was not limiting the transport and biodegradation of contaminants under the experimental conditions. This is also observed in the simulated pore water concentration profiles (not shown) found to be similar to the bulk water simulated concentration profiles presented in Figures 7.2, 7.3 and 7.4. This suggests the biofilm is fully penetrated (i.e. no substrate limitation and all reactions take place over the full depth of the biofilm), which may have allowed the omission of diffusion limitations in some previous modeling work for subsurface flow constructed wetland biofilms (García et al., 2010). The other factors influencing substrate transport, i.e. the biofilm thickness and the liquid boundary layer (through which transport from the bulk water to the biofilm surface occurs by molecular diffusion) had a moderate to significant sensitivity (Table 7.1). The sensitivity of the temperature (not shown) ranked high among the parameters with a "strong effect" on the bulk liquid concentrations only at the higher temperatures studied (i.e. at 20°C and 24°C).

There were no steep spatial gradients of the biomass profiles inside the biofilm within the simulation time (not shown), an expected result for a thin biofilm, whereas the simulation time scale (20 d) may also have been too short for the development of significant changes in bacterial species distribution (Vanhooren, 2002; Samsó and García, 2013). According to our current knowledge, a period of 3 year of continuous wetland operation is about sufficient for the bacterial communities to stabilize (Samsó and García, 2013). The apparent homogeneous distribution of the bacterial species involved can be of advantage for the processes, as microorganisms with differing redox potential requirements reside in close proximity, making the exchange of intermediate products between the species more efficient.

Anaerobic species favored by the oxygen limited conditions within the wetland mesocosm dominated the biofilm. The biomass volume fractions show the biofilm developed essentially as an anaerobic biofilm with a significant community of sulphate reducing and methanogenic bacteria (Table 7.3). Their population was enhanced with increase in temperature (Table 7.3), in agreement with the observation that in most cases the growth rates of both methanogens and sulphate reducing bacteria increase with increasing temperature (Baptista et al., 2003), while the methanogenic species were found to dominate the activity within the biofilm across all temperatures during the simulation period, with activity defined here as the product of all substrate quotients in the Monod growth equation for a given population (Shanahan and Semmens, 2004).

7.4   Conclusions

Based on the CWM1 biokinetic model and the 1-D biofilm model of the AQUASIM software, the dynamics of biofilm growth in a subsurface constructed wetland mesocosm have been simulated. The in silico analysis of constructed wetland biofilms indicates that aerobic, anoxic and anaerobic active biomass develops in the shallow biofilm. The development of this "active biomass", and thus its effect on the biofilm growth was influenced by the availability of substrates, presence or absence of macrophytes and temperature. Anaerobic biomass dominated the biofilm with methanogenic activity being the main organic matter removal process. For operational analysis and constructed wetland technology development, a complex model such as the CWM1 model is recommended, with further extensions as required to address factors such as biofilm and rhizosphere development dynamics. The CWM1-Aquasim-Biofilm model is a useful tool to show the influence of the rhizosphere configuration on the performance of the constructed wetlands.

7.5   References

Baptista, J.D.C., T. Donnelly, D. Rayne and R.J. Davenport, 2003. Microbial mechanisms of carbon removal in subsurface flow wetlands. Water Science and Technology, 48: 127-134.

Boltz, J.P., E. Morgenroth, D. Brockmann, C. Bott, W.J. Gellner and P.A. Vanrolleghem, 2011. Systematic evaluation of biofilm models for engineering practice: components and critical assumptions. Water Science and Technology, 64: 930-944.

Brockmann, D., K.H. Rosenwinkel and E. Morgenroth, 2008. Practical identifiability of biokinetic parameters of a model describing two-step nitrification in biofilms. Biotechnology and Bioengineering.

Caselles-Osorio, A., A. Porta, M. Porras and J. García, 2007. Effect of High Organic Loading Rates of Particulate and Dissolved Organic Matter on the Efficiency of Shallow Experimental Horizontal Subsurface-flow Constructed Wetlands. Water, Air, & Soil Pollution, 183: 367-375.

Faulwetter, J.L., V. Gagnon, C. Sundberg, F. Chazarenc, M.D. Burr, J. Brisson, A.K. Camper and O.R. Stein, 2009. Microbial processes influencing performance of treatment wetlands: A review. Ecological Engineering, 35: 987-1004.

Gagnon, V., F. Chazarenc, Y. Comeau and J. Brisson, 2007. Influence of macrophyte species on microbial density and activity in constructed wetlands. Water Science and Technology, 56: 249-254.

García, J., D.P.L. Rousseau, J. MoratÓ, E.L.S. Lesage, V. Matamoros and J.M. Bayona, 2010. Contaminant Removal Processes in Subsurface-Flow Constructed Wetlands: A Review. Critical Reviews in Environmental Science and Technology, 40: 561-661.

Haberl, R., 1999. Constructed wetlands: A chance to solve wastewater problems in developing countries. Water Science and Technology, 40: 11-17.

Henze, M., W. Gujer, T. Mino and M. van Loosdrecht, 2000. Activated sludge models ASM1, ASM2, ASM2d and ASM3. Scientific and Technical Report No. 9. . IWA Publishing, London, UK.

Kadlec and S. Wallace, 2009. Treatment wetlands. 2nd ed. Boca Raton, Fla: CRC Press, 1048 pp.

Krasnits, E., E. Friedler, I. Sabbah, M. Beliavski, S. Tarre and M. Green, 2009. Spatial distribution of major microbial groups in a well established constructed wetland treating municipal wastewater. Ecological Engineering, 35: 1085-1089.

Kumar, J.L.G. and Y.Q. Zhao, 2011. A review on numerous modeling approaches for effective, economical and ecological treatment wetlands. Journal of Environmental Management, 92: 400-406.

Langergraber, G., 2001. Development of a simulation tool for subsurface flow constructed wetlands. . *Wiener Mitteilungen* 169, Vienna, Austria, 207p. ISBN 3-85234-060-8.

Langergraber, G., D.P.L. Rousseau, J. Garcia and J. Mena, 2009. CWM1: a general model to describe biokinetic processes in subsurface flow constructed wetlands. Water Science and Technology, 59: 1687-1697.

Langergraber, G. and J. Šimůnek, 2005. Modeling Variably Saturated Water Flow and Multicomponent Reactive Transport in Constructed Wetlands. Vadose Zone Journal, 4: 924-938.

Langergraber, G. and J. Šimůnek, 2012. Reactive Transport Modeling of Subsurface Flow Constructed Wetlands Using the HYDRUS Wetland Module. Vadose Zone Journal, 11.

Lazarova, V. and J. Manem, 1995. Biofilm characterization and activity analysis in water and wastewater treatment. Water Research, 29: 2227-2245.

Llorens, M.W. Saaltink, M. Poch and J. García, 2011. Bacterial transformation and biodegradation processes simulation in horizontal subsurface flow constructed wetlands using CWM1-RETRASO. Bioresource Technology, 102: 928-936.

Mayo, A.W. and T. Bigambo, 2005. Nitrogen transformation in horizontal subsurface flow constructed wetlands I: Model development. Physics and Chemistry of the Earth, Parts A/B/C, 30: 658-667.

Mburu, N., D. Sanchez-Ramos, Diederik P.L. Rousseau, J.J.A van Bruggen, George Thumbi, Otto R. Stein, Paul B. Hook and P. N.L.Lens, 2012. Simulation of carbon, nitrogen and sulphur conversion in batch-operated experimental wetland mesocosms. Ecological Engineering.

Mburu, N., S. Tebitendwa, D. Rousseau, J. van Bruggen and P. Lens, 2013. Performance Evaluation of Horizontal Subsurface Flow–Constructed Wetlands for the Treatment of Domestic Wastewater in the Tropics. Journal of Environmental Engineering, 139: 358-367.

McBride, G.B. and C.C. Tanner, 1999. Modelling biofilm nitrogen transformations in constructed wetland mesocosms with fluctuating water levels. Ecological Engineering, 14: 93-106.

Melo, L.F., 2005. Biofilm physical structure, internal diffusivity and tortuosity. Water Science & Technology, Vol 52: 77–84.

Munch, C., P. Kuschk and I. Roske, 2005. Root stimulated nitrogen removal: only a local effect or important for water treatment? Water Sci Technol, 51: 185-192.

Neralla, S., R.W. Weaver, B.J. Lesikar and R.A. Persyn, 2000. Improvement of domestic wastewater quality by subsurface flow constructed wetlands. Bioresource Technology, 75: 19-25.

Ojeda, E., J. Caldentey, M.W. Saaltink and J. Garcia, 2008. Evaluation of relative importance of different microbial reactions on organic matter removal in horizontal subsurface-flow constructed wetlands using a 2D simulation model. Ecological Engineering, 34: 65-75.

Petersen, B., K. Gernaey and P.A. Vanrolleghem, 2001. Practical identifiability of model parameters by combined respirometric-titrimetric measurements. Water Science and Technology, 43: 347-355.

Reichert, 1998. AQUASIM 2.0 - User Manual Computer Program for the Identi cation and Simulation of Aquatic Systems. Swiss Federal Institute for Environmental Science and Technology (EAWAG) CH - 8600 Dubendorf Switzerland.

Reichert, P., 1995. Design techniques of a computer program for the identification of processes and the simulation of water quality in aquatic systems. Environmental Software, 10: 199-210.

Rousseau, D.P.L., 2005. Performance of constructed treatment wetlands: model-based evaluation and impact of operation and maintenance. PhD Thesis, Ghent University, Ghent, Belgium (available from http://biomath.ugent.be/publications/download/).

Rousseau, D.P.L., P.A. Vanrolleghem and N. De Pauw, 2004. Model-based design of horizontal subsurface flow constructed treatment wetlands: a review. Water Research, 38: 1484-1493.

Samsó, R. and J. Garcia, 2013. BIO_PORE, a mathematical model to simulate biofilm growth and water quality improvement in porous media: Application and calibration for constructed wetlands. Ecological Engineering, 54: 116-127.

Şeker, Ş., H. Beyenal and A. Tanyolaç, 1995. The effects of biofilm thickness on biofilm density and substrate consumption rate in a differential fluidizied bed biofilm reactor (DFBBR). Journal of Biotechnology, 41: 39-47.

Shanahan, J.W. and M.J. Semmens, 2004. Multipopulation Model of Membrane-Aerated Biofilms. Environmental Science & Technology, 38: 3176-3183.

Truu, M., J. Juhanson and J. Truu, 2009. Microbial biomass, activity and community composition in constructed wetlands. Science of the Total Environment, 407: 3958-3971.

Vanhooren, H., 2002. Modelling for optimisation of biofilm wastewater treatment processes: A complexity compromise. PhD Thesis, Ghent University, Ghent, Belgium.

Vymazal, J., 2010. Constructed Wetlands for Wastewater Treatment: Five Decades of Experience†. Environmental Science & Technology, 45: 61-69.

Wanner, O. and E. Morgenroth, 2004. Biofilm modeling with AQUASIM. Water Sci Technol, 49: 137-144.

Wichern, M., C. Lindenblatt, M. Lübken and H. Horn, 2008. Experimental results and mathematical modelling of an autotrophic and heterotrophic biofilm in a sand filter treating landfill leachate and municipal wastewater. Water Research, 42: 3899-3909.

Wik, T., 1999. On Modeling the Dynamics of Fixed Biofilm Reactors with focus on nitrifying trickling filters. PhD Thesis. Chalmers University of Technology Goteborg, Sweden.

Wynn, T.M. and S.K. Liehr, 2001. Development of a constructed subsurface-flow wetland simulation model. Ecological Engineering, 16: 519-536.

Zhang, C.-B., J. Wang, W.-L. Liu, S.-X. Zhu, H.-L. Ge, S.X. Chang, J. Chang and Y. Ge, 2010. Effects of plant diversity on microbial biomass and community metabolic profiles in a full-scale constructed wetland. Ecological Engineering, 36: 62-68.

Zysset, A., F. Stauffer and T. Dracos, 1994. Modeling of reactive groundwater transport governed by biodegradation. Water Resour. Res., 30: 2423-2434.

Chapter 8:   General discussion and outlook

8.1   Introduction

Constructed wetlands (CWs) are a natural alternative to technical methods of wastewater treatment (Hijosa-Valsero et al. 2012; Stottmeister et al. 2003). They are the engineer-made equivalent of natural wetlands designed to reproduce and intensify the wastewater treatment processes that occur in natural wetlands (IWA 2000; Mara 2004). Through the efforts of research and operation for over fifty years, CWs have now been successfully used for environmental pollution control, through the treatment of a wide variety of wastewaters including industrial effluents, urban and agricultural stormwater runoff, animal wastewaters, leachates, sludges and mine drainage (Babatunde et al. 2008; IWA 2000). The main attraction for using constructed wetlands in pollution control has not only been due to their functional values, but also because of their low investment scale, running and maintenance costs compared to the traditional sewage treatment processes (Haberl 1999; Kivaisi 2001; Okurut 2000).

In spite of its relative simplicity of construction, operation and maintenance, the successful design and optimization of CWs remains challenging. The various treatment processes taking place within the CW have not been entirely understood, quantified and integrated into the design models for CWs. This claim is indirectly supported by the fact that literature contains many studies aiming at contributing to the expert design and further understanding of CW processes (Chapter 1). The information does not translate well to simple design rules. Thus, there still remains a lack of integrated knowledge or models due to the complexity of the CW system. This for instance, is unlike for the mature technologies like activated sludge process or the anaerobic treatment of domestic wastewater, where modeling is an established method for assessing wastewater treatment for design, systems analysis, operational analysis, and control.

The effectiveness of different types of wetland vegetation, environmental conditions, microbial dynamics (colonization characteristics of certain groups of microorganisms), wastewater composition, filter material and loading rates have been identified as important driving factors influencing the performance of the CWs. Their

163

design is a function of the wastewater nature, its pollutant load, the available area to build the wetland and the climatic conditions of the site (Hijosa-Valsero et al. 2012). However, the continuous development and consolidation in the understanding of these factors has challenged the state-of -the art design of CWs, which is based on rules of thumb, empirical and steady-state equations (Kadlec 2000; Rousseau et al. 2004; Stein et al. 2006). At the same time, the accumulated scientific information and engineering experiences about the CWs processes has seen the development of a number of processes based models describing these systems (Langergraber and Šimůnek 2012). These mechanistic models that describe the transformation and elimination processes taking place within CWs have become a promising tool for understanding parallel processes and interactions occurring in wetlands (Llorens et al. 2011a; Ojeda et al. 2008).

It is thus clear that the continuous development and application of CW knowledge is a key step to improve the design and optimization (most efficient configuration) of constructed wetlands. On the other hand, the achievement of a design model with universal parameters for the different CWs systems is unlikely. The different set of conditions for each CW system under study and the complex treatment processes represent additional significant challenges to the designer. Yet, CWs are anticipated to be used more especially in the tropical regions for wastewater treatment.

8.2   Horizontal Subsurface Flow Constructed Wetland

The research carried out in this thesis focused on the understanding of the performance and treatment processes in horizontal subsurface flow constructed wetlands (HSSF-CW) with respect to carbon, nitrogen and sulphur conversions. Over the last decade, the CW wastewater technology has evolved into new reactor configurations featuring different flow characteristics, and a much broader range of treatment applications. It has been observed that the configuration and nature of a CW affects its performance (García et al. 2004). In particular, the HSSF-CW constitutes an appropriate technology for treating wastewaters that have been subjected to primary clarification. The use of HSSF-CWs is especially well suited to the removal of suspended solids, organic matter and nitrogen (Caselles-Osorio et al. 2007). HSSF-CWs have complex processes driven by plants, microorganisms, soil matrixes and substances in wastewater interacting with each other. These constituents influence the purification of influent wastewater as well as the overall quality of the effluent (Allen et al. 2002; Kadlec and Wallace 2009; Thurston et al. 2001). Thus, the interpretation of experimental measurements and data through a dynamic process based model approach remains indispensable to obtain insights of the different pollutant removal processes involved in wastewater treatment within HSSF-CW systems.

Currently, CW mechanistic mathematical models are used to determine the relationships between the different biogeochemical processes and weigh their relative contributions in wastewater treatment (Langergraber and Šimůnek 2012; Llorens et al. 2011b; Mburu et al. 2012; Ojeda et al. 2008). Where the availability of experimental data/measurements about the system behaviour is limited e.g. plant root oxygen release, inverse modeling with mechanistic models has been found useful (Mburu et al. 2012). In 2009, a general model named the Constructed Wetland Model No. 1 (CWM1) was published with the aim of providing a widely accepted model formulation for biochemical transformation and degradation processes of organic matter, nitrogen and sulphur in subsurface flow constructed wetlands (SSF CWs) (Langergraber et al. 2009). Indeed, the introduction of the CWM1 is aimed at the dynamic modeling of the HSSF-CWs and eventually the optimization of design and operation of HSSF-CWs for cost reduction and improvement in effluent quality. Further, CWM1 is intended to introduce common terminology for modeling HSSF-CWs processes. This is important as it has not been always clear how discovered knowledge and information is integrated or transferred to the existing pool of CWs design knowledge. Thus, similarly as for the activated sludge models which has proved useful for facilitating the transfer of knowledge between researchers, for instance, by using a compact matrix formulation for the mathematical models, CWM1 is intended to introduce common terminology for modeling HSSF-CWs processes.

Thus, the main body of the thesis was focused on the study of SSF CWs in two parts:

- Pilot scale experimental study
- Mechanistic simulations of reactive transport and organic matter transformation and degradation

165

8.2.1 Pilot scale experimental study

Undertaking a pilot scale study on a tropical HSSF-CW treating domestic wastewater was in an attempt to provide performance data that can guide design and operation under tropical conditions. The pilot scale studies were carried out at Juja (Kenya) with three gravel based HSSF-CW cells. Primary effluent of domestic wastewater diverted from the outflow of the primary facultative pond of the Jomo-Kenyatta University of Agriculture and Technology sewage treatment works was used as the influent (with a continuous flow) to the wetlands. As described in Chapter 3, the three HSSF-CW cells (each of 7.5 m x 3 m x 0.5 m) out of which two were planted with the local emergent macrophyte *Cyperus papyrus*, were monitored. Influent-effluent water quality data were collected (which included sulphate measurements as a possible alternative electron acceptor that contributes to the oxidation of organic matter in HSSF-CW). Through statistical analysis, trends of variables and significant correlations were looked for among operating and environmental conditions as well as the performance of the constructed wetland. A literature survey in Chapter 2 on the application of the tropical *Cyperus papyrus* macrophyte in constructed wetlands was used to deduce its potential contribution to wastewater treatment. The plants' involvement in the input of oxygen into the root zone, in the uptake of nutrients, the generation of biomass and its harvesting, water loss (transpiration), among others, were discussed. Further, a comparative evaluation of the performance and economics of the pilot HSSF-CW and an operational waste stabilization pond (WSP) system was undertaken (chapter 4). This was to offer technical and economic insights that would simplify technology selection among the two non-conventional sanitation technologies with potential for wide application in the developing countries.

8.2.2 Mechanistic simulations of reactive transport and organic matter transformation and degradation

Process based numerical simulations of reactive transport and organic matter transformation and degradation in subsurface flow constructed wetlands was done within the framework of the Constructed Wetland Model No.1 (CWM1) biokinetic model. The data used were obtained from the pilot HSSF-CW and a separate investigation with batch-operated constructed wetland mesocosms. Simulation results are presented for the pilot HSSF-CW (Chapter 5) and the wetland mesocosms (Chapters 6 and 7). The recent developments in numerical mechanistic modeling for subsurface flow CWs have served as an important, low-cost tool for a better description and an improved understanding of the internal functioning of CW systems. Together with this, model results are only useful if the model predictions are reliable and transferable. Hence, before the models can be applied as tools to refine wetland design criteria and operational modes, there is a need to apply the models to a wide range of data sets to test their predictive power, reduce parameter uncertainty by calibration, modify process equations or extend the models with relevant processes. This can be achieved with lab-scale, pilot scale or full scale experiments.

8.3   Evaluation of treatment performance of pilot HSSF-CW

Chapter 3 revealed the successful performance of the wetland cells in terms of compliance with local discharge standards as stipulated by NEMA-Kenya (2003) with respect to COD, BOD_5, TSS and SO_4^{2-}-S at an average mass removal efficiency between 58.9% and 74.9%. The removal efficiency of the pilot CWs cells was assessed based on the data from a sampling campaign period of about 2 years (Table 7.2 of chapter 7). However, CWs are intended to treat wastewater during decades. Therefore, the application of those data obtained during a pilot trial period to the design and/or maintenance of full scale CWs can result in an unexpected decrease in the long-term system performance (Hijosa-Valsero et al. 2012). Reduced oxygen effluent concentrations of less than 2 mg/l and a well-buffered pH of between 7 - 7.5, characterized the vegetated wetland cells (Table 7.2 of chapter 7). A large proportion of the COD and BOD_5 removal probably occurred by anoxic and anaerobic processes, among others denitrification and sulphate reduction. This was supported by the observation that nitrates did not accumulate (notwithstanding the modest nitrification rates) and the significant sulphate removal in the pilot scale HSSF-CW cells (Chapter 3). In the control cell, the curve for mass loading rate versus removal rates for COD and TSS (Figures 3.4 and 3.5 of chapter 3 respectively) had a brief range of linear correlation compared to that of the planted cells, depicting the planted cells as being capable of sustaining higher loading and removal rates. This was linked to the presence of the macrophyte *Cyperus papyrus* and the spread of its roots and rhizomes, possibly promoting physical and biological removal processes associated with bulk

167

pollutants. The observed rate of BOD decay of 0.1 m d^{-1} (Chapter 3) indicates that the HSSF-CWs are area requirement competitive when compared to the widely applied waste stabilization pond system in the tropics.

Removal of nutrients, nitrogen and phosphorus, from wastewater by the HSSF-CW cells was consistently not satisfactory during the monitoring period (Chapter 3). Low oxygen concentrations that prevailed in the wetland cells limited possible ammonium removal via the nitrification-denitrification pathway. Further, the nitrogen removal via plant uptake did not translate to substantial removal performance (Chapter 3). Despite the significant difference in the phosphorus removal between the planted and the unplanted cells, phosphorus removal via routes other than plant uptake was limited in the gravel bed HSSF-CW systems. This is attributable to the low phosphorus retention capacity by the granitic gravel used in the HSSF-CW bed. Overall, the results showed that achieving good and simultaneous reduction of bulk pollutants and nutrients under similar operating conditions was not feasible. Therefore, it is recommended that the intended objective of treatment be clearly defined at the wetland design stage in order to avoid over expectations of treatment performance of the system (Okurut 2000).

Chapter 3 showed the technical viability of using HSSF-CW within the tropical environments with respect to the removal of COD, BOD, and TSS. Furthermore, the side by side economic and performance assessment of the pilot constructed wetland and a waste stabilization pond in the treatment of domestic wastewater augmented this potential (Chapter 4). The main attraction for using constructed wetlands has not only been promoted because of their technical functionality, but also because of the favorable economic cost of setting them up. The deduction from this comparison is that constructed wetlands can be established competitively with waste stabilization ponds in the tropical environments (Chapter 4).

8.4   Processes based simulation of HSSF-CW

The simulations were conducted in the Constructed Wetland Model No.1 (CWM1) framework that is based on the mathematical formulation as introduced by the IWA Activated Sludge Models (ASMs). CWM1 describes the biochemical transformation and degradation processes for organic matter, nitrogen and sulphur in subsurface flow constructed wetlands, including hydrolysis, aerobic respiration, nitrification and denitrification, sulphate reduction and methanogenesis (Langergraber et al. 2009). The main objective of CWM1 is to predict effluent concentrations from constructed wetlands without predicting gaseous emissions.

For purposes of simulations, the CWM1 biokinetic model was applied as CWM1-RETRASO model, as implemented in the RetrasoCodeBright (RCB) by Llorens et al (2012) (Chapter 5) and as CWM1-AQUASIM model as implemented in the simulation software for identification and simulation for aquatic systems, AQUASIM by Mburu et al. (2012) (Chapter 6 and 7).

8.4.1   Simulation with CWM1-RETRASO model

This was applied for the reactive transport simulation of the pilot scale HSSF-CW (chapter 5). The RCB code provides the knowledge related to reactive transport and flow properties (Llorens et al. 2011a). The pilot-scale HSSF-CW reactive transport calibrations and validations were performed against the observed concentrations of COD, NH_4^+-N, NO_3^--N and SO_4^{2-}-S in the effluent over five different influent wastewater flow rates and compositions (Chapter 5). This involved the estimation of the initial bacterial biomass concentration required as input to run the CWM1-RETRASO model. Results of the reactive transport simulations showed the distribution of the biodegradation pathways and wastewater components within the HSSF-CW (Figures 5.3, 5.4, 5.5 and 5.6 of Chapter 5). Anaerobic processes occurred over larger areas of the simulated wetland. The dissolved oxygen concentration was low in the influent and dropped rapidly within the first meter of the simulated HSSF-CW. The simulation results of the aerobic processes distribution showed their location to be close to the oxygen sources. Aerobic respiration in the simulated wetland was only observed in a thin layer of water on the surface of the wetland and at the inlet. The hydrolysis processes were found to take place mainly near the inlet zone of the simulated HSSF-CW. The exclusion of the plant processes in the CWM1-RETRASO model should not be a major limitation when only considering domestic wastewater, due to their relative low uptake rate of nutrients compared to the conversion and elimination rates caused by micro-organisms (Langergraber 2001).

8.4.2 Simulation with CWM1-AQUASIM model

This was executed in AQUASIM's mixed reactor compartment without biofilm as CWM1-AQUASIM (Chapter 6) and with biofilm consideration as CWM1-AQUASIM-Biofilm (Chapter 7). Data from 16 subsurface-flow wetland mesocosms operated under controlled greenhouse conditions with three different plant species (*Typha latifolia, Carex rostrata, Schoenoplectus acutus*) and an unplanted control were used. The experimental constructed wetlands were operated under controlled greenhouse conditions at Montana State University in Bozeman (Montana, USA). Details of column design, construction and planting, as well as sampling and measurement are fully described in Allen et al. (2002) and Stein et al. (2006). Further, by compiling the macrophyte process (growth, root oxygen release, physical degradation, decay and oxygen leaching), physical re-aeration, as well as reverse sorption processes for COD and ammonium in AQUASIM (Chapter 6), the combined effect of availability of electron acceptors, temperature and macrophyte type was studied.

Figures 6.2, 6.3 and 6.4 of Chapter 6 show that the bulk water of the constructed wetland mesocosms was well simulated with the CWM1-AQUASIM. A sensitivity analysis performed for the CWM1-AQUASIM showed that in general the yield coefficients for methanogenic, sulphate reducing, fermenting and heterotrophic bacteria; the saturation/inhibition coefficients for oxygen, sulphate, acetate, fermentable COD and hydrolysis; the plant growth rate constant, rate constant for lysis of heterotrophic bacteria and the COD sorption coefficients are among the most sensitive and distinct parameters affecting the predicted concentrations (Table 6.1 of chapter 6). The simulation results showed that the measured behavior of the batch operated subsurface flow system could only be modeled well when COD and ammonium sorption are considered as additional process. Plant processes (root oxygen leaching, uptake of ammonium, physical degradation and decay) did not have much impact on the treatment compared to the microbiological processes.

In simulations with the CWM1-AQUASIM-Biofilm model, the biofilm density and microbial volume fractions were the most sensitive biofilm characteristics for the majority of the variables in the bulk liquid zone, while diffusion coefficients of components into the biofilm had a low sensitivity to the final simulation results at all temperatures. Simulation results showed the development of a thin biofilm dominated by anaerobic bacteria (fermenting bacteria, acetotrophic sulphate reducing bacteria and acetotrophic methanogenic bacteria). This correlated well with the observed and evaluated degradation activities and inferred redox conditions within the HSSF-CW bed (Chapter 6). The biofilm density and microbial concentration were evaluated to be influenced by temperature and the wetland vegetation. The consideration of biofilm processes was found to be only important qualitatively in defining and providing insights to the microbial dynamics and biofilm characteristics in the constructed wetland mesocosms (Chapter 7).

In both evaluations, i.e. with and without biofilm consideration, anaerobic microbial degradation pathways were found to dominate compared to the anoxic and aerobic pathways on the account of limited oxygen supply and renewal in the subsurface flow constructed wetlands. Indeed, sulphate reduction and methanogenesis were simulated to be the more widespread degradation reaction. Thus, subsurface constructed wetlands with minimal oxygen renewal capacities such as the horizontal subsurface flow wetlands should be designed as an anaerobic or an aerobic–anaerobic hybrid system (Cui et al. 2006), rather than as an aerobic system.

8.5   Simulation platform

Computer aided simulations are fastening the development of constructed wetland models. The biokinetic model CWM1 was implemented in the two software platforms AQUASIM and RCB. The AQUASIM software provided the possibility of using the inbuilt automatic sensitivity analysis and parameter estimation features, which were used to optimize and fit parameters for the CWM1 biokinetic model, plant processes, sorption and the biofilm compartment (Chapter 6). Further, AQUASIM features a biofilm reactor compartment that describes the growth and population dynamics of biofilms in which substrate gradients over the depth are important. Data treatment in the software was relatively straightforward. However, batch simulation (in the mixed reactor compartment of AQUASIM) has the advantage of simplifying the hydraulics when integrating transport and transformation processes in porous media that are otherwise solved with advection-dispersion-reaction equations in 1D or 2D. On the other hand, the RetrasoCodeBright provided a platform to undertake modeling in 2D.

Both the calibration for the biochemical and flow model, i.e. considering the residence time distribution was possible (Chapter 5). Data treatment with the software was intricate. The code does not have neither a graphical user interface (GUI) nor parameter estimation tools. Running the 2D simulation was time consuming. The RCB code architecture does not consider fixed biomass and therefore does not allow to conduct biofilm based-modeling.

Overall AQUASIM afforded the freedom and flexibility of specifying the CWM1 biokinetic model in the software, which is essential to eliminate barriers for potential model users, while producing realistic simulation results. Despite this, the CWM1 model does not generally use first order kinetics but rather Monod kinetics, which captures more knowledge, but introduces model complexity because parameters for constructed wetland processes have not been well estimated or standardised and assessment of parameter variability is limited. Thus, the calibration and application of this model would require a high-quality data set with a high information content. These are generally rare and could be developed in future dedicated studies to couple CW performance evaluation with design modeling.

8.6   Outlook

Over the years, numerous examples have shown that the constructed wetland wastewater treatment technology is suitable for treating both municipal and a broad range of industrial wastewaters (Kadlec and Wallace 2009; Vymazal and Kröpfelová 2009). Originally, merely small wetlands were constructed which could only treat the wastewater produced by small populations. In the meantime, larger treatment works have also been built that are able to cope with a population equivalent of several thousands (Stottmeister et al. 2003). Therefore, their mechanistic simulation will continue to augment the much needed insights and information on the working mechanisms of CW systems.

The improved understanding of the treatment processes in the "black box" constructed wetland acquired through the process modeling of CWs will ensure wastewater is treated as efficiently as possible. With the good performance for removal of the wastewater pollutants, and the unraveling of the internal treatment processes, the constructed wetland wastewater treatment technology will be adapted with as much confidence in their operability and pollutant removal levels as with comparable conventional wastewater treatment technology. This will see an increased institutional and public support and application of this inexpensive green technology of wastewater management, especially in the tropical developing countries.

The testing of the CWM1 biokinetic model framework with diverse data and the development of techniques to directly measure or verify parameters of CWM1 and other mechanistic processes considered for CWs is desirable. This will enhance the calibration and further application of these mechanistic models for evaluation and refining of CW design criteria, as it is the case for other technologies, such as the activated sludge processes. Improvement in parameter certainty may see the extension of the model to include other relevant processes of CW (e.g. clogging) and the possible simplification of the model for standard design simulations. The evolution of methodologies to measure kinetic and stoichiometric parameters; to characterization wastewater (Ortigara et al. 2011) and to determine the microbial composition (DeJournett et al. 2007) in constructed wetlands is an important step in this direction. Further, significant insight may be gained from the recently developed approaches and techniques in related fields such as contaminant hydrology, environmental microbiology and biotechnology which can enhance mechanistic investigations and open up new possibilities for process characterization and interpretation of constructed wetland performance. Achieving a better understanding of the complex interactions involved will enable the basic scientific aspects to be optimally combined with the technical possibilities available, thus enabling wetland technologies to be used on a broader scale.

Providing environmentally-safe sanitation to millions of people remains a significant challenge, especially in the developing countries. The target for goal 7 of the Millennium Development Goals (MDGs) calls on countries to halve, by 2015, the proportion of people without improved sanitation facilities (from 1990 levels). Constructed wetlands which have contributed to providing low cost sanitation in the developed temperate countries will prove helpful on a large scale for the developing countries as a sustainable alternative to the large distance, waterborne conveyance and high-energy input wastewater treatment systems. It is hoped this work on subsurface wetland treatment performance evaluation and mechanistic simulation has augmented this outlook.

8.7   Conclusions

- Constructed wetland wastewater treatment technology has developed in importance, mainly due to the need for low cost, effective wastewater treatment. In particular, the CW technology is likely to find wider applications in the developing tropical countries.

- The main current motivations for modelling CW processes remain as (a) the acquisition of design parameters and laws, (b) the optimisation and the prediction of CW performance and (c) technology development for the long-term sustainability of sewage treatment.

- The CWMI was found adept in simulating the putative in-situ biotic degradation pathways and providing insight into the complex and heterogeneous constructed wetland system.

8.8   References

Allen, W. C., et al. (2002), 'Temperature and wetland plant species effects on wastewater treatment and root zone oxidation', Journal of Environmental Quality, 31 (3), 1010-16.

Babatunde, A. O., et al. (2008), 'Constructed wetlands for environmental pollution control: A review of developments, research and practice in Ireland', Environment International, 34 (1), 116-26.

Caselles-Osorio, Aracelly, et al. (2007), 'Effect of High Organic Loading Rates of Particulate and Dissolved Organic Matter on the Efficiency of Shallow Experimental Horizontal Subsurface-flow Constructed Wetlands', Water, Air, & Soil Pollution, 183 (1), 367-75.

Cui, L. H., et al. (2006), 'Performance of hybrid constructed wetland systems for treating septic tank effluent', J Environ Sci, 18 (4), 665-9.

DeJournett, Todd D., Arnold, William A., and LaPara, Timothy M. (2007), 'The characterization and quantification of methanotrophic bacterial populations in constructed wetland sediments using PCR targeting 16S rRNA gene fragments', Applied Soil Ecology, 35 (3), 648-59.

García, J., et al. (2004), 'Initial contaminant removal performance factors in horizontal flow reed beds used for treating urban wastewater', Water Research, 38 (7), 1669-78.

Haberl, R. (1999), 'Constructed wetlands: A chance to solve wastewater problems in developing countries', Water Science and Technology, 40 (3), 11-17.

Hijosa-Valsero, M., Sidrach-Cardona, R., and Bécares, E. (2012), 'Comparison of interannual removal variation of various constructed wetland types', Science of the Total Environment, 430, 174-83.

IWA (2000), 'Constructed wetlands for pollution control. Processes, performance, design and operation', Scientific and Technical report No.8.

Kadlec (2000), 'The inadequacy of first-order treatment wetland models', Ecological Engineering, 15 (1-2), 105-19.

Kadlec and Wallace, S. (2009), 'Treatment wetlands', 2nd ed. Boca Raton, Fla: CRC Press, 1048 pp.

Kivaisi, Amelia K. (2001), 'The potential for constructed wetlands for wastewater treatment and reuse in developing countries: a review', Ecological Engineering, 16 (4), 545-60.

Langergraber, G. (2001), 'Development of a simulation tool for subsurface flow constructed wetlands. ', Wiener Mitteilungen 169, Vienna, Austria, 207p. ISBN 3-85234-060-8.

Langergraber, G. and Šimůnek, Jirka (2012), 'Reactive Transport Modeling of Subsurface Flow Constructed Wetlands Using the HYDRUS Wetland Module', Vadose Zone Journal, 11 (2).

Langergraber, G., et al. (2009), 'CWM1: a general model to describe biokinetic processes in subsurface flow constructed wetlands', Water Science and Technology, 59 (9), 1687-97.

Llorens, Saaltink, Maarten W., and García, Joan (2011a), 'CWM1 implementation in RetrasoCodeBright: First results using horizontal subsurface flow constructed wetland data', Chemical Engineering Journal, 166 (1), 224-32.

Llorens, et al. (2011b), 'Bacterial transformation and biodegradation processes simulation in horizontal subsurface flow constructed wetlands using CWM1-RETRASO', Bioresource Technology, 102 (2), 928-36.

Mara, D. (2004), 'Domestic Wastewater Treatment in Developing Countries', Earthscan UK and US. Cromwell Press, Trowbridge, UK.

Mburu, N, et al. (2012), 'Simulation of carbon, nitrogen and sulphur conversion in batch-operated experimental wetland mesocosms', Ecological Engineering.

Ojeda, E., et al. (2008), 'Evaluation of relative importance of different microbial reactions on organic matter removal in horizontal subsurface-flow constructed wetlands using a 2D simulation model', Ecological Engineering, 34 (1), 65-75.

Okurut, T. O. (2000), 'A Pilot Study on Municipal Wastewater Treatment Using a Constructed Wetland in Uganda', PhD dissertation, UNESCO-IHE, Institute for Water Education, Delft, The Netherlands.

Ortigara, A. R., Foladori, P., and Andreottola, G. (2011), 'Kinetics of heterotrophic biomass and storage mechanism in wetland cores measured by respirometry', Water Sci Technol, 64 (2), 409-15.

Rousseau, D. P. L., Vanrolleghem, P. A., and De Pauw, N. (2004), 'Model-based design of horizontal subsurface flow constructed treatment wetlands: a review', Water Research, 38 (6), 1484-93.

Stein, Otto R., et al. (2006), 'Plant species and temperature effects on the k-C* first-order model for COD removal in batch-loaded SSF wetlands', Ecological Engineering, 26 (2), 100-12.
Stottmeister, U., et al. (2003), 'Effects of plants and microorganisms in constructed wetlands for wastewater treatment', Biotechnology Advances, 22 (1-2), 93-117.

Thurston, Jeanette A., et al. (2001), 'Fate of indicator microorganisms, giardia and cryptosporidium in subsurface flow constructed wetlands', Water Research, 35 (6), 1547-51.

Vymazal and Kröpfelová, L (2009), 'Removal of organics in constructed wetlands with horizontal sub-surface flow: A review of the field experience', Science of the Total Environment, 407 (13), 3911-22.

Summary

Sustainable sanitation and water pollution control calls for adoption of affordable and efficient wastewater treatment technologies. In the developing countries, characterized as they are by inadequate sanitation, the safe management of wastewater is not widespread. There is therefore a need for an appropriate technology that can reliably achieve acceptable effluent quality for discharge to the environment at minimal cost. Constructed wetland (CW) systems have been used as a cost effective alternative to conventional methods of wastewater treatment. However, the mechanistic understanding of the CW has not matured, while performance data that can guide design and operation of CW under tropical climate are scarce.

This study was undertaken to explore the treatment of domestic wastewater with subsurface constructed wetlands, in order: 1) to provide performance data that can influence design and operation of CW under tropical conditions and, 2) to evaluate the processes involved with the transformation and degradation of organic matter and nutrients.

In the study a pilot scale horizontal subsurface flow constructed wetland (HSSF-CW) was established in Kenya, and an existing dataset with experimental data from 16 wetland mescocosms operated under greenhouse conditions were obtained from the U.S.A., kindly shared by Montana State University. The data from the tropical pilot scale HSSF-CW was used to conduct a performance evaluation (with respect to organic matter, nutrients and suspended solids) (Chapter 3), a comparison (performance and economics) with a waste stabilization pond (Chapter 4) and a reactive transport simulation in the HSSF-CW (Chapter 5). The data from the wetland mescocosms were used to conduct a simulation of carbon, nitrogen and sulphur conversion in the batch-operated experimental wetland mesocosms with and without consideration of biofilm development within the wetland (Chapters 6 and 7).

The pilot HSSF-CW consisted of three cells receiving a continuous feed of primary effluent from the outflow of a primary facultative pond. In two of the cells, the macrophyte *Cyperus papyrus* was planted, while the third cell acted as a control. The wetland cells were 7.5 m long and 3 m wide with vertical masonry sides, 0.95 m deep, and a concrete floor sloped at one percent. The cells were filled with granite type gravel to a depth of 0.6 m, ranging in size from 9-37 mm, with a porosity of 45 %.The experimental constructed wetlands were operated under controlled greenhouse conditions at Montana State University in Bozeman (Montana, USA). 16 subsurface constructed wetland mesocosms were constructed from polyvinyl chloride (PVC) pipes (60 cm in height $\times$ 20 cm in diameter) and filled to a depth of 50 cm with washed pea-gravel (0.3–1.3 cm in diameter). Four columns each were planted with *Carex utriculata, Schoenoplectus acutus* and *Typha latifolia*, while four were left unplanted as controls. A series of 20 days incubations with artificial wastewater was conducted over 20 months at temperatures ranging from 4 to 24 ∘C at 4 ∘C steps. A synthetic wastewater simulating secondary domestic effluent was used with mean influent concentrations of 470 mg/l COD, 44 mg/l N (27 Org- N, 17 NH_4^+-N), 8 mg/l PO_4^{3-}-P, and 14 mg/l SO_4^{2-}-S. Columns were gravity drained 3 days prior to each incubation and then again at the start of each incubation. Upon each emptying, columns were refilled from above with new wastewater. Sampling from all 16 columns occurred at days 0, 1, 3, 6, 9, 14 and 20 of each incubation and those sub-samples were analysed afterwards for the constituents.

The results of the study showed successful performance of the tropical HSSF-CW for the secondary treatment of domestic wastewater with respect to organic matter (BOD_5 and COD) and TSS removal at an average mass removal efficiency between 58.9 and 74.9 %. Moderate (36 - 49 %) removal of nutrients (nitrogen and phosphorus) was recorded. A two days hydraulic retention time was found to be optimum for organic matter removal. The presence of the macrophyte enhanced the ability of the wetland to withstand higher organic and suspended solids loading. The land area requirement for secondary treatment (based on BOD_5 removal) was estimated as 2.0 m^2 per population equivalent (Chapter 3). A waste stabilization pond would need 3 times the area that would be required for the HSSF-CW to treat the same amount of wastewater under tropical conditions (chapter 4). The evaluation of the capital cost of HSSF-CW system showed that it is largely influenced by the size of the population served, local cost of land and the construction materials involved.

Using a (mechanistic) numerical model that incorporates the growth of six microbial groups (heterotrophic, autotrophic nitrifying, fermenting, acetotrophic methanogenic, acetotrophic sulphate reducing and the sulphide oxidising bacteria) and the subsequent consumption of electron donors and acceptors, the influence of key operating and environmental conditions on the biochemical transformation and degradation processes for organic matter, nitrogen and sulphur in subsurface flow constructed wetlands was evaluated (Chapters 5, 6 and 7). Sorption processes were found to be important in simulating COD and ammonia removal in subsurface flow constructed wetlands. The rates of oxygen transfer by physical re-aeration and root oxygen transfer were found insufficient, indicating that organic matter in the wastewater was removed mainly by anaerobic processes. Indeed anaerobic reactions occurred over large areas of the simulated HSSF-CW and contributed (on average) to the majority (68%) of the COD removal, compared to aerobic (38%) and anoxic (1%) reactions in the tropical HSSF-CW (Chapter 5). Further a thin (predominantly anaerobic) biofilm in which no concentration gradients developed was simulated. The simulations suggest that incorporation of plant roots into the substrate of constructed wetlands enhances microbial populations related to the transformation and degradation of pollutants in constructed wetlands. Measured and simulated data demonstrate that the resultant effect on the wetland performance may not necessarily be related to temperature.

This research contributed to performance data and getting a better mechanistic understanding about the factors influencing the performance of horizontal subsurface flow constructed wetland treating real domestic wastewater under tropical conditions. The findings obtained in this research may prove useful towards the wider application of the constructed wetland wastewater treatment technology and the optimization of full-scale HSSF-CW.

Samenvatting

Duurzame sanitatie en het beheersen van waterverontreiniging pleiten voor het gebruik van betaalbare en efficiënte technologieën voor afvalwaterzuivering. In ontwikkelingslanden, getypeerd door onvoldoende sanitatie, is het veilig beheer van afvalwater niet wijdverbreid. Er is daarom behoefte aan een passende technologie die op een betrouwbare manier een acceptabele lozingskwaliteit van het effluent kan bereiken tegen minimale kosten. Helofytenfilter systemen (CW) zijn reeds in gebruik als een kosteneffectief alternatief voor conventionele afvalwaterzuiveringsmethoden. Ons begrip van de interne processen in CW blijft echter beperkt, en gegevens die kunnen leiden tot een beter ontwerp en werking van CW in een tropische klimaat blijven schaars.

Dit onderzoek werd uitgevoerd om de zuivering van huishoudelijk afvalwater met ondergronds doorstroomde helofytenfilters beter te doorgronden, met de volgende doelstellingen: 1) prestatie gegevens verzamelen die van invloed kunnen zijn op het ontwerp en de werking van CW onder tropische omstandigheden en, 2) de processen evalueren die betrokken zijn bij de transformatie- en afbraakprocessen van organische stof en nutriënten.

Voor dit onderzoeksproject werd een ondergronds horizontaal doorstroomde helofytenfilter (HSSF-CW) als proefproject gebouwd in Kenia, en werd ook gebruik gemaakt van een bestaande set experimentele gegevens van 16 mesocosmos-opstellingen uit een gesloten kasexperiment in de VS die ter beschikking gesteld werden door de Montana State University. De gegevens van het proefproject werden gebruikt om zuiveringsrendementen na te gaan (met betrekking tot organische stof, nutriënten en zwevende stoffen) (hoofdstuk 3), om een vergelijking te maken (zuiveringsrendement en kosten) met een stabilisatievijver (hoofdstuk 4) en om een reactief transport simulatie uit te voeren (Hoofdstuk 5). De gegevens van de mesocosmos-opstellingen werden gebruikt om een simulatie van de koolstof, stikstof en zwavel omzettingen uit te voeren, met en zonder inachtneming van biofilm ontwikkeling binnen de mesocosmossen (hoofdstukken 6 en 7).

De helofytenfilter bestond uit drie cellen die op continue wijze gevoed werden met primair gezuiverd effluent afkomstig van de uitlaat van een primaire facultatieve stabilisatievijver. Twee van de cellen werden beplant met de macrofyt Cyperus papyrus, terwijl de derde cel als onbeplante controle fungeerde. Elke cel was 7,5 m lang en 3 m breed met gemetselde opstaande wanden, 0,95 m diep en met een cementen vloer met een helling van 1 procent. De cellen werden tot een diepte van 0,6 m gevuld met een laag granitisch grind met een deeltjesdiameter tussen 9-37 mm en een porositeit van 45%. De experimenten met de mesocosmos-opstellingen vonden plaats in een kas met klimaatcontrole, bij Montana State University in Bozeman (Montana, VS). Zestien mesocosmossen werden gebouwd van PVC buizen (60 cm hoog x 20 cm diameter) en gevuld met een laag gewassen parelgrind van 50 cm (0,3 – 1,3 cm diameter). Telkens vier kolommen werden beplant met respectievelijk Carex utriculata, Schoenoplectus acutus en Typha latifolia, terwijl de resterende 4 kolommen als onbeplante controle fungeerden. Een 20-maanden durend experiment werd opgezet waarbij de temperatuur in stappen van 4°C aangepast werd tussen 4 °C en 24 °C. De kolommen werden om de 23 dagen gevuld met artificieel afvalwater met gemiddelde influentconcentraties van 470 mg/l COD, 44 mg/l N (27 Org- N, 17 NH_4^+-N), 8 mg/l PO_4^{3-}-P, en 14 mg/l SO_4^{2-}-S; tussen de verschillende cycli werden de kolommen gedurende 3 dagen gedraineerd. Watermonsters werden genomen op dag 0, 1, 3, 6, 9, 14 en 20 van iedere cyclus en geanalyseerd op COD, NH_4^+ en SO_4^{2-}.

De resultaten van dit onderzoek betreffende het gebruik van tropische HSSF-CW voor de secundaire zuivering van huishoudelijk afvalwater hebben een goede prestatie aangetoond voor wat betreft organisch materiaal (BOD5 en COD) en zwevende stoffen met een gemiddelde massa verwijdering tussen 58,9 en 74,9%. Voor nutriënten (stikstof en fosfor) werd daarentegen slechts een matige (36-49%) verwijdering opgetekend. Een hydraulische verblijftijd van 2 dagen bleek optimaal voor de verwijdering van organisch materiaal. De aanwezigheid van planten verbeterde het vermogen van de helofytenfilter om met hogere belastingen van organische en zwevende stoffen om te gaan.

De vereiste oppervlakte voor secundaire zuivering (gebaseerd op BOD5 verwijdering) werd ingeschat op 2,0 m2 per inwoner equivalent (hoofdstuk 3). Een stabilisatievijver zou een 3x grotere oppervlakte nodig hebben dan een HSSF-CW om dezelfde hoeveelheid afvalwater te zuiveren onder tropische condities (hoofdstuk 4). Een evaluatie van de investeringskosten voor een HSSF-CW systeem toonde aan deze kosten sterk beïnvloed worden door het aantal aangesloten personen, de plaatselijke grondprijzen en de gebruikte bouwmaterialen.

Door middel van een (mechanistisch) numeriek model, met daarin opgenomen de groei van zes groepen bacteriën (heterotrofe, autotrofe nitrificerende, fermenterende, acetotrofe methanogene, zwavel reducerende en sulfide oxiderende bacteriën) en het corresponderende gebruik van verschillende elektronendonoren en –acceptoren, werd de invloed van een aantal belangrijke operationele en omgevingsparameters op de biochemische transformatieprocessen van organisch materiaal, stikstof en zwavel in ondergronds doorstroomde helofytenfilters geëvalueerd (hoofdstukken 5, 6 en 7). Een belangrijke vaststelling was dat sorptieprocessen een belangrijke rol speelden bij het correct simuleren van COD en ammonium processen. Verder bleken fysische en biologische (via de plantenwortels) zuurstoftransfersnelheden ontoereikend, wat erop wijst dat organisch materiaal in het afvalwater voornamelijk via anaerobe processen verwijderd wordt. Uit de simulaties bleek inderdaad dat in tropische HSSF-CW de anaerobe processen in grote delen van de helofytenfilter voorkwamen en gemiddeld gezien bijdroegen tot het grootste deel (68%) van de COD verwijdering, in vergelijking met aerobe (38%) en anoxische (1%) processen (hoofdstuk 5). In een volgende simulatie werd een dunne (hoofdzakelijk anaerobe) biofilm gebruikt zonder concentratiegradiënten. De resultaten hiervan suggereren dat de plantenwortels in het substraat van helofytenfilters de microbiële populaties stimuleren die verantwoordelijk zijn voor de omzetting en afbraak van polluenten. De gemeten zowel als gesimuleerde waarden toonden ook aan dat zuiveringsrendementen niet noodzakelijk gerelateerd zijn aan de temperatuur.

Ter conclusie kan gesteld worden dat dit onderzoek een bijdrage geleverd heeft aan de databank van zuiveringsrendementen en vooral ook aan de kennis over de verschillende processen en factoren die hierop een invloed hebben, vooral dan in ondergronds horizontaal doorstroomde helofytenfilters voor de zuivering van huishoudelijk afvalwater onder tropische condities. De bekomen resultaten kunnen nuttig blijken om de bredere toepassing van helofytenfilters te stimuleren en om hun werking te optimaliseren.

About the Author

Njenga Mburu holds an MSc. and a BSc. degree in Civil Engineering obtained, respectively, in 1997 and 2004, from the Jomo Kenyatta University of Agriculture and Technology (JKUAT), Juja, Kenya. He is a member of the Engineers Board of Kenya (EBK), the Institute of Engineers of Kenya (IEK), and a licensed lead Expert with the National Environmental Management Authority in Kenya.

After his undergraduate studies, he worked as a hydrologist for the Ministry of Water in Kenya. Between 2002-2004, in the course of his post-graduate studies at JKUAT, he served as a part-time lecturer at the department of Civil and Environmental Engineering and was also engaged as a training logistics officer at the African Institute for Capacity Development (AICAD), Juja, Kenya.

Njenga's research focuses on performance evaluation and mechanistic simulation of the subsurface constructed wetland for wastewater treatment. His current research interests include sustainable sanitation concepts and appropriate pollution prevention technologies for the developing countries.

Njenga Mburu is currently a lecturer in the department of Civil and Structural Engineering at the Masinde Muliro University of Science and Technology, Kakamega, Kenya. He teaches both undergraduate and post-graduate courses in water and wastewater treatment, and public health engineering.

Netherlands Research School for the
Socio-Economic and Natural Sciences of the Environment

C E R T I F I C A T E

The Netherlands Research School for the
Socio-Economic and Natural Sciences of the Environment
(SENSE), declares that

Njenga Mburu

born on 26 March 1973 in Mombasa, Kenya

has successfully fulfilled all requirements of the
Educational Programme of SENSE.

Delft, 29 November 2013

the Chairman of the SENSE board

Prof. dr. Rik Leemans

the SENSE Director of Education

Dr. Ad van Dommelen

The SENSE Research School has been accredited by the Royal Netherlands Academy of Arts and Sciences (KNAW)

K O N I N K L I J K E   N E D E R L A N D S E
A K A D E M I E   V A N   W E T E N S C H A P P E N

The SENSE Research School declares that Mr. Njenga Mburu has successfully fulfilled all requirements of the Educational PhD Programme of SENSE with a work load of 45 ECTS, including the following activities:

SENSE PhD Courses
o   Environmental Research in Context
o   Research Context Activity: Communicating PhD research process and outcome to a general audience of stakeholders

Other PhD and Advanced MSc Courses
o   Horizontal flow constructed wetlands modelling using the CWM1-RETRASO software
o   Wetlands for Water Quality
o   Wastewater Treatment, Design and Modelling

Management and Didactic Skills Training
o   Thesis supervision of MSc student with thesis title: A Comparison of the Performance and Economics of a Waste Stabilization Pond and Constructed Wetland System in Juja, Kenya

Oral Presentations
o   *Modelling flow and solute transport in a pilot-scale horizontal Subsurface Flow Constructed Wetland.* Wetland Pollutant Dynamics and Control (WETPOL) Symposium, 20-24 September, 2009, Barcelona, Spain
o   *Performance evaluation and 2-D mechanistic simulation of a tropical HSSF-CW treating domestic wastewater in Kenya.* 12th International Conference on Wetland Systems for Water Pollution Control, 4-9 October, 2010, Venice, Italy
o   *Potential of Cyperus payrus macrophyte for wastewater treatment: A review.* 12th International Conference on Wetland Systems for Water Pollution Control, 4-9 October, 2010, Venice, Italy

SENSE Coordinator PhD Education

Dr. ing. Monique Gulickx

T - #0424 - 101024 - C192 - 240/170/10 - PB - 9781138015524 - Gloss Lamination